APPROXIMATION OF SET-VALUED FUNCTIONS

Adaptation of Classical Approximation Operators

APPROXIMATION OF SET-VALUED FUNCTIONS

Adaptation of Classical Approximation Operators

Nira Dyn
Elza Farkhi
Alona Mokhov

Tel Aviv University, Israel

Imperial College Press

ICP

Published by

Imperial College Press
57 Shelton Street
Covent Garden
London WC2H 9HE

Distributed by

World Scientific Publishing Co. Pte. Ltd.
5 Toh Tuck Link, Singapore 596224
USA office: 27 Warren Street, Suite 401-402, Hackensack, NJ 07601
UK office: 57 Shelton Street, Covent Garden, London WC2H 9HE

Library of Congress Cataloging-in-Publication Data
Dyn, N. (Nira), author.
   Approximation of set-valued functions : adaptation of classical approximation operators /
Nira Dyn, Tel Aviv University, Israel, Elza Farkhi, Tel Aviv University, Israel, Alona Mokhov,
Tel Aviv University, Israel.
     pages cm
   Includes bibliographical references and index.
   ISBN 978-1-78326-302-8 (hardcover : alk. paper)
   1. Approximation theory. 2. Linear operators. 3. Function spaces. I. Farkhi, Elza, author.
II. Mokhov, Alona, author. III. Title.
   QA221.D94 2014
   515'.8--dc23

                                                                                    2014023451

British Library Cataloguing-in-Publication Data
A catalogue record for this book is available from the British Library.

Typeset by Stallion Press
Email: enquiries@stallionpress.com

Preface

This book is concerned with the approximation of set-valued functions. It mainly presents our work on the design and analysis of approximation methods for functions mapping the points of a closed real interval to *general compact* sets in $\mathbb{R}^n$. Most previous results on approximation of set-valued functions were confined to the special case of functions with compact convex sets in $\mathbb{R}^n$ as their values.

We present approximation methods together with bounds on the approximation error, measured in the Hausdorff metric. The error bounds are given in terms of the regularity of the approximated set-valued function. The regularity properties used are mainly of low order, such as Hölder continuity and bounded variation. This facilitates the analysis of approximation methods for non-smooth set-valued functions, which are common in areas such as optimization and control. The obtained error estimates are of similar quality to those for real-valued functions.

Our work was motivated by the need to approximate a set-valued function from a finite number of its samples. Such a need arises in the problem of "reconstruction" of a 3D object from its parallel cross-sections, which are compact 2D sets, and also in the numerical solution of non-linear differential inclusions. In the latter problem the set-valued solution has to be approximated from a discrete collection of its computed values, which are not necessarily convex sets.

The approach taken in this book is to adapt classical linear approximation operators for real-valued functions to set-valued functions. For sample-based operators, the main method of adaptation is to replace operations between numbers by operations between sets. The main difficulty in this approach is the design of set operations, which yield operators with approximation properties. A second method is based on representations of set-valued functions by collections of real-valued functions. Having such a representation at hand, the approximation of the set-valued function is reduced to the approximation of the corresponding collection of representing real-valued functions. The main effort in this approach is the design of an appropriate representation consisting of real-valued functions with regularity properties "inherited" from those of the approximated set-valued function.

The book consists of three parts. The first presents basic notions and results needed to establish the adapted approximation methods, and to carry out their analysis. The second part is concerned with several approximation methods for set-valued functions with compact sets in $\mathbb{R}^n$ as their values. The third part is devoted to the simpler case $n = 1$, where special representations of such set-valued functions are designed, and approximation methods based on these representations are discussed.

The subject of the book is on the border of the two fields Set-Valued Analysis and Approximation Theory. The panoramic view, given in the book can attract researchers from both fields to this intriguing subject. In addition, the book will be useful for researchers working in related fields such as control and game theory, mathematical economics, optimization and geometric modeling.

The bibliography covers various related topics. To improve the readability of the book, references to the bibliography do not appear in the text, but are deferred to special sections, mostly at the end of chapters.

We would like to thank the School of Mathematical Sciences at Tel-Aviv University for giving us a supporting and stimulating environment for carrying out our research, and for presenting it in this book.

Tel-Aviv, May 2013
The authors

Contents

III Approximation of SVFs with Images in $\mathbb{R}$ 107

Notations

PART I

Scientific Background

Chapter 1

On Functions with Values
in Metric Spaces

This chapter introduces basic notions, definitions and notation related to univariate functions with values in metric spaces. In particular we recall two basic approximations for such functions, the piecewise constant and piecewise linear interpolants.

For real-valued functions we recall known facts on approximation by sample-based operators. This is the main type of operators which are adapted to set-valued functions in the book. Here we also review spline subdivision schemes, which are adapted to sets in forthcoming chapters.

Throughout the book, Schoenberg spline operators, Bernstein polynomial operators and polynomial interpolation operators are used to demonstrate the various adaptations of approximation operators defined on real-valued functions to set-valued functions. In this chapter we recall the definition of these operators and known error estimates. For Schoenberg and Bernstein operators we also recall stable evaluation algorithms based on repeated binary averages.

1.1. Basic Notions

This section discusses the notions of continuity, Hölder continuity and bounded variation for functions defined on $[a, b]$ with values in a metric space (X, ρ). In this book ρ is either the Euclidean distance in $\mathbb{R}^n$ or the Hausdorff distance between subsets of $\mathbb{R}^n$.

We denote by $\chi = \{x_0, \ldots, x_N\}$, $a = x_0 < x_1 < \cdots < x_N = b$ a partition of $[a, b]$, with maximal step size $|\chi| = \max\{x_{i+1} - x_i : i = 0, \ldots, N - 1\}$. The uniform partition is denoted by χ_N where $x_i = a + ih$, $i = 0, \ldots, N$, and $h = |\chi_N| = (b - a)/N$.

The notions of modulus of continuity and of total variation are central to the approximation of set-valued functions, and therefore we recall them below.

The modulus of continuity of $f \colon [a, b] \to X$ is

$$\omega_{[a,b]}(f, \delta) = \sup\{\, \rho(f(x), f(y)) :$$
$$|x - y| \leq \delta,\ x, y \in [a, b]\,\}, \quad \delta \in (0, b - a]. \qquad (1.1)$$

We call f bounded if $\omega_{[a,b]}(f, b - a) < \infty$. As for real-valued functions the following property of $\omega_{[a,b]}(f, \delta)$

$$\omega_{[a,b]}(f, \lambda\delta) \leq (1 + \lambda)\,\omega_{[a,b]}(f, \delta), \quad \lambda, \delta \in \mathbb{R}_+. \qquad (1.2)$$

holds in metric spaces too. Obviously, if f is continuous on $[a, b]$ then $\omega_{[a,b]}(f, \delta) \to 0$ as $\delta \to 0$. The collection of all continuous functions on $[a, b]$ is denoted by $C[a, b]$. A function $f \in C[a, b]$ is Hölder continuous with exponent $\nu \in (0, 1]$ (or shortly Hölder-ν), if

$$\rho(f(x), f(z)) \leq C|x - z|^{\nu}, \quad x, z \in [a, b],$$

with C a constant depending only on f and $[a, b]$.

Among the Hölder continuous functions we denote by $Lip([a, b], L)$ the collection of all Lipschitz continuous functions satisfying

$$\rho(f(x), f(y)) \leq L|x - y|, \quad \forall\, x, y \in [a, b].$$

The variation of $f : [a, b] \to X$ on a partition $\chi = \{x_0, \ldots, x_N\}$,

$$V(f, \chi) = \sum_{i=1}^{N} \rho(f(x_i), f(x_{i-1})).$$

The total variation of f on $[a, b]$ is defined by

$$V_a^b(f) = \sup_{\chi} V(f, \chi).$$

A function f is of bounded variation if $V_a^b(f) < \infty$. For f of bounded variation, we consider the variation function

$$v_f(x) = V_a^x(f), \quad x \in [a, b]. \qquad (1.3)$$

It is easy to see that v_f is non-decreasing. Moreover the following claim is known.

Proposition 1.1.1 *A function $f : [a, b] \to X$ is continuous and of bounded variation on $[a, b]$ if and only if v_f is a continuous function on $[a, b]$.*

Proof The sufficiency follows from

$$\rho(f(x), f(y)) \leq V_x^y(f) = v_f(y) - v_f(x), \quad \text{for } x < y. \tag{1.4}$$

To prove the other direction, fix $x \in [a, b]$ and $\varepsilon > 0$. By the uniform continuity of f on $[a, b]$, $\rho(f(z), f(y)) < \varepsilon/2$ whenever $|z - y| < \delta$ for some $\delta > 0$. First we show that v_f is continuous from the left. We can always choose $\chi = \{x_0, \dots, x_N\}$, with $a = x_0$, $x_N = x$, such that $x - x_{N-1} < \delta$ and

$$V_a^x(f) < V(f, \chi) + \varepsilon/2 = \sum_{i=1}^{N} \rho(f(x_i), f(x_{i-1})) + \varepsilon/2.$$

Thus

$$V_a^x(f) < \sum_{i=1}^{N-1} \rho(f(x_i), f(x_{i-1})) + \varepsilon,$$

implying that $v_f(x) - v_f(x_{N-1}) < \varepsilon$. By the monotonicity of v_f we get for every $x_{N-1} < y < x$

$$v_f(x) - v_f(y) < \varepsilon.$$

Similarly one can show the continuity of v_f from the right. Thus we obtain that v_f is continuous at x and consequently it is continuous on $[a, b]$. $\square$

It is easy to conclude from (1.4) that

$$\omega_{[a,b]}(f, \delta) \leq \omega_{[a,b]}(v_f, \delta). \tag{1.5}$$

We denote the set of all functions $f : [a, b] \to X$ which are continuous and of bounded variation by $CBV[a, b]$. We call $f \in CBV[a, b]$ a **CBV function**.

For single-valued functions mapping $[a, b]$ into $\mathbb{R}^n$ it is possible to define moduli of smoothness:

$$\omega_{k,[a,b]}(f, \delta) = \sup_{|h| \leq \delta} \{|\Delta_h^k f(x)| : x, x + kh \in [a, b]\},$$

where $\Delta_h^k f(x) = \sum_{i=0}^{k} \binom{k}{i} (-1)^{k-i} f(x + ih)$, $k \geq 1$ and $|\cdot|$ is the Euclidean norm. Clearly

$$\omega_{1,[a,b]}(f, \delta) = \omega_{[a,b]}(f, \delta),$$

and

$$\omega_{k,[a,b]}(f, \delta) \leq 2\omega_{k-1,[a,b]}(f, \delta). \tag{1.6}$$

The known relation

$$\omega_{k,[a,b]}(f, \delta) \leq \delta\omega_{k-1,[a,b]}(f', \delta)$$

implies for a function with Hölder-ν continuous derivative of order $r \leq k - 1$, that

$$\omega_{k,[a,b]}(f, \delta) = O(\delta^{r+\nu}). \tag{1.7}$$

We denote by $C^k[a, b]$ the collection of functions mapping $[a, b]$ into $\mathbb{R}^n$, with continuous k-th derivative, and say that f is C^k if it belongs to $C^k[a, b]$. For $f : [a, b] \to \mathbb{R}^n$ we define the norm $\|f\|_\infty = \sup_{x \in [a,b]} |f(x)|$.

1.2. Basic Approximation Methods

First we consider piecewise constant interpolants, which are well defined in metric spaces.

Given a metric space (X, ρ) and $f : [a, b] \to X$, this approximant is defined relative to a partition χ of $[a, b]$, $a = x_0 < x_1 < \cdots < x_N = b$, by

$$\begin{aligned} S_\chi^0 f(x) &= f(x_i), \quad x \in [x_i, x_{i+1}), \quad i = 0, \ldots, N - 1, \\ S_\chi^0 f(x_N) &= f(x_N). \end{aligned} \tag{1.8}$$

The error of this method is bounded by

$$\rho(f(x), S_\chi^0 f(x)) \leq \omega_{[a,b]}(f, |\chi|), \quad x \in [a, b]. \tag{1.9}$$

In particular, for Hölder-ν functions

$$\rho(f(x), S_\chi^0 f(x)) \leq C_\nu |\chi|^\nu.$$

Note that $S_\chi^0 f$ is a discontinuous function interpolating f at the points of χ.

In geodesic metric spaces it is possible to construct a continuous interpolant. First we define an averaging operation based on a geodesic.

For given $x, y \in X$, let $\gamma(\cdot) \subset X$ be a geodesic curve connecting x and y, such that $\gamma(0) = x$, $\gamma(\rho(x, y)) = y$, and

$$\rho(x, y) = \rho(x, \gamma(s)) + \rho(\gamma(s), y), \quad 0 \le s \le \rho(x, y).$$

The t-weighted geodesic average of x, y depending on the choice of γ is

$$tx \oplus (1 - t)y = \gamma((1 - t)\rho(x, y)), \quad t \in [0, 1]. \tag{1.10}$$

This implies the metric property

$$\rho(tx \oplus (1 - t)y, sx \oplus (1 - s)y) = |t - s|\rho(x, y), \quad t, s \in [0, 1],$$

which in its weaker form is

$$\rho(tx \oplus (1 - t)y, x) = (1 - t)\rho(x, y), \quad t \in [0, 1]. \tag{1.11}$$

With this operation one can define a continuous piecewise linear interpolant given by

$$S_\chi^1 f(x) = \frac{x_{i+1} - x}{x_{i+1} - x_i} f(x_i) \oplus \frac{x - x_i}{x_{i+1} - x_i} f(x_{i+1}), \quad x \in [x_i, x_{i+1}], \tag{1.12}$$

for $i = 0, \dots, N - 1$.

A bound on the error of this approximation is easily derived from the triangle inequality and (1.11),

$$\rho(f(x), S_\chi^1 f(x)) \le \frac{3}{2}\omega_{[a,b]}(f, |\chi|), \quad x \in [a, b]. \tag{1.13}$$

Indeed, for $|x - x_i| \le \frac{1}{2}|\chi|$,

$$\rho(f(x), S_\chi^1 f(x)) \le \rho(f(x), f(x_i)) + \rho(f(x_i), S_\chi^1 f(x))$$

$$\le \omega_{[a,b]}(f, |\chi|) + \frac{1}{2}\omega_{[a,b]}(f, |\chi|).$$

1.3. Classical Approximation Operators

Here we give a short overview of the approximation procedures for real-valued functions that we adapt to the set-valued case.

We consider approximating operators for real-valued functions on $[a, b]$, and focus mostly on sample-based operators of the form

$$A_\chi f(x) = \sum_{i=0}^{N} c_i(x) f(x_i). \tag{1.14}$$

All operators considered in this book have error estimates of the form

$$|A^\delta f(x) - f(x)| \le C\omega_{[a,b]}(f, \phi(x, \delta)), \tag{1.15}$$

where A^δ is an operator depending on a positive parameter $\delta > 0$ and where $\phi : [a, b] \times \mathbb{R}_+ \to \mathbb{R}_+$ is a continuous real-valued function, non-decreasing in its second argument, with $\phi(x, 0) = 0$. When the operator depends on a partition χ, $\delta = |\chi|$.

The general form (1.15) corresponds to many approximation operators, approximating continuous real-valued functions. For many of them $\phi(x, \delta) = \delta$. Yet for the important case of the Bernstein operators $\phi(x, \delta) = \sqrt{x(1-x)\delta}$.

We restrict the class (1.14) by requiring

$$\sum_{i=0}^{N} c_i(x) = 1, \quad x \in [a, b], \tag{1.16}$$

a condition which is necessary for (1.15), since the modulus of continuity of any constant function is zero.

Remark 1.3.1 For single-valued functions $f : [a, b] \to \mathbb{R}^n$, the operator (1.14) is well defined. In this case the absolute value in the left-hand side of the estimate (1.15) denotes the Euclidean vector norm.

In Chapter 6 we use representations of sample-based operators in terms of repeated binary weighted averages. Such a representation exists for any operator of the form (1.14) satisfying (1.16), but is not unique. Yet, for the concrete positive operators that we study, stable evaluation procedures based on such a representation are well known, and are recalled below.

1.3.1. *Positive operators*

Here we state a well-known approximation result on general positive linear operators and discuss some specific classical operators.

A linear operator A is called positive if, for any non-negative f, $Af(x)$ is non-negative. Obviously, A_χ in (1.14) is a positive linear operator if

$c_i(x) \geq 0$, $i = 0, \ldots, N$, $x \in [a, b]$. It is a global operator if $c_i(x) > 0$, $x \in (a, b)$, $i = 0, \ldots, N$, and it is a local operator if for all i, $c_i(x) > 0$ only in a neighborhood of x_i, and is zero otherwise.

The following result is a quantitative form of the Korovkin theorem due to Freud.

Theorem 1.3.2 *Let A be a positive linear operator defined on continuous real-valued functions and denote $e_i(x) = x^i, i = 0, 1, 2$. If*

$$\max_{i=0,1,2} \|Ae_i - e_i\|_\infty \leq \lambda, \tag{1.17}$$

then there is a constant C such that for any $f \in C[a, b]$

$$|Af(x) - f(x)| \leq C(\lambda \|f\|_\infty + \omega_{2,[a,b]}(f, \sqrt{\lambda})), \quad x \in [a, b]. \tag{1.18}$$

It follows from this theorem and (1.6), (1.7) that for Hölder-ν f, the error is $O(\lambda^{\frac{\nu}{2}})$, while for f with Hölder-ν first derivative the error is $O(\lambda^{\frac{\nu+1}{2}})$.

As concrete operators we consider the Bernstein polynomial operators and the Schoenberg spline operators defined on $[0, 1]$. For a general interval $[a, b]$ the corresponding operators are defined by the linear transformation $t = a + (b - a)x$, $x \in [0, 1]$. In this case the sampling points are

$$x_i = a + ih, \quad i = 0, 1, \ldots, N, \quad h = \frac{b - a}{N},$$

and in the error estimates $\frac{1}{N}$ is replaced by h. To introduce these operators we need the notation $\mathcal{P}_n$ for the space of polynomials of degree up to n.

Bernstein polynomial operators

The Bernstein polynomial operator B_N for a real-valued function $f : [0, 1] \to \mathbb{R}$ is a classical global operator given by

$$(B_N f)(x) = \sum_{i=0}^{N} \binom{N}{i} x^i (1 - x)^{N-i} f\left(\frac{i}{N}\right) \in \mathcal{P}_N, \quad x \in [0, 1]. \tag{1.19}$$

The value $B_N f$ can be calculated recursively by repeated binary averages, using the de Casteljau algorithm,

$$f_i^0 = f(i/N), \ i = 0, \ldots, N,$$

$$f_i^k = (1 - x)f_i^{k-1} + xf_{i+1}^{k-1}, \quad i = 0, 1, \ldots, N - k, \ k = 1, \ldots, N, \tag{1.20}$$
$$B_N f(x) = f_0^N.$$

This evaluation procedure is illustrated by the diagram in Fig. 1.3.3.

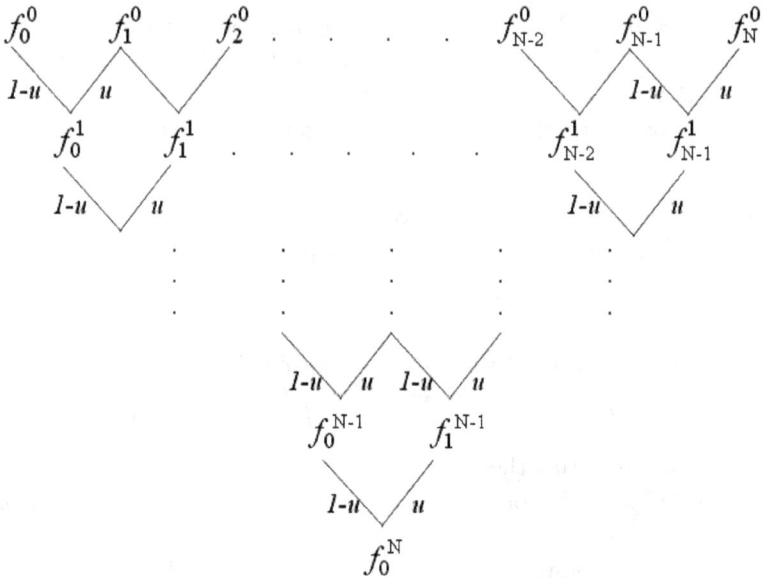

Figure 1.3.3   Evaluation of $B_N f(u)$ by the de Casteljau algorithm.

This stable algorithm is an important tool in Computer-Aided Geometric Design.

It is known that there exists a constant C independent of f and x such that for a continuous f

$$|f(x) - B_N f(x)| \le C\omega_{[0,1]}(f, \sqrt{x(1-x)/N}), \quad x \in [0,1]. \tag{1.21}$$

Remark 1.3.4 It follows from (1.21) that if f is Hölder-ν, then the error of Bernstein approximation is

$$|f(x) - B_N f(x)| \le CN^{-\frac{\nu}{2}}, \quad x \in [0,1], \tag{1.22}$$

with C a generic constant depending on f.

A refined estimate for smooth functions may be obtained from (1.18) and (1.7). It is well-known that for the Bernstein operator λ in (1.17) is $O(\frac{1}{N})$. Thus, if f has a Hölder-ν first derivative, then

$$|f(x) - B_N f(x)| \le CN^{-\frac{\nu+1}{2}}, \tag{1.23}$$

with C as above.

Schoenberg spline operators

For $f \in C[0,1]$ the Schoenberg spline operator is a typical example of a local operator defined as

$$S_{m,N}f(x) = \sum_{i=0}^{N} f\left(\frac{i}{N}\right) b_m(Nx - i), \tag{1.24}$$

where b_m denotes the B-spline of order m (degree $m-1$) with integer knots, namely $b_m x \mid_{[i,i+1]} \in \mathcal{P}_{m-1}$, $b_m(x) \in C^{m-2}(\mathbb{R})$ and $\mathrm{supp}\,(b_m) = [0,m]$. The set $\{b_m(x-i),\ i \in \mathbb{Z}\}$ is a basis of the space of spline functions (splines) of order m (degree $m-1$) with integer knots. Moreover,

$$b_m(x) > 0,\ x \in (0,m), \quad \sum_{i \in \mathbb{Z}} b_m(\cdot - i) \equiv 1. \tag{1.25}$$

It is a known approximation result that

$$|S_{m,N}f(x) - f(x)| \le \left\lfloor \frac{m+1}{2} \right\rfloor \omega_{[a,b]}\left(f, \frac{1}{N}\right), \quad x \in \left[\frac{m-1}{N}, 1\right], \tag{1.26}$$

with $\lfloor x \rfloor$ the maximal integer not greater than x.

Schoenberg spline operators $S_{m,N}f$ can be evaluated by an algorithm, based on the recurrence formula for B-splines. This is known as the de Boor algorithm.

For $\frac{m-1}{N} \le x < 1$, let $l \in \{m-1, m, m+1, \ldots, N-1\}$ be such that $\frac{l}{N} \le x < \frac{l+1}{N}$, and let $u = Nx$. Obviously $l \le u \le l+1$. $S_{m,N}f(x)$ is calculated by

$$f_i^0 = f\left(\frac{i}{N}\right), \quad i = l - m + 1, \ldots, l.$$

For $k = 1, \ldots, m-1$

$$f_i^k = \frac{i+m-k-u}{m-k} f_{i-1}^{k-1} + \frac{u-i}{m-k} f_i^{k-1},$$

$$i = l - m + k + 1, \ldots, l. \tag{1.27}$$

$$S_{m,N}f(x) = f_l^{m-1}.$$

Observe that f_i^k is a convex combination of f_{i-1}^{k-1} and f_i^{k-1}, which guarantees the stability of the algorithm.

The de Boor algorithm can be represented by a diagram similar to that in Fig. 1.3.3 with the weights of the convex combinations not fixed but changing according to (1.27).

Remark 1.3.5 The symmetric Schoenberg spline operator for m even is

$$\widehat{S}_{m,N}f(x) = \sum_{i=0}^{N} f\left(\frac{i}{N}\right) \widehat{b}_m(Nx - i), \quad \widehat{b}_m(x) = b_m\left(x - \frac{m}{2}\right). \quad (1.28)$$

For $x \in \left[\frac{i}{N}, \frac{i+1}{N}\right)$, $\widehat{S}_{m,N}f(x)$ in (1.28) is a convex combination of values of f at a set of symmetric points relative to $\left(\frac{i}{N}, \frac{i+1}{N}\right)$. The evaluation of $\widehat{S}_{m,N}f$ is similar to that of $S_{m,N}f$.

The approximation of a function with Hölder-ν first derivative is $O(N^{-1-\nu})$, with the constant in the estimate depending on f only through its Hölder constant.

1.3.2. *Interpolation operators*

For a real-valued function $f \in C[a, b]$ the polynomial interpolation operator at the set of points χ is

$$P_\chi f(x) = \sum_{i=0}^{N} l_i(x)f(x_i), \quad \text{with } l_i(x) = \prod_{j=0, j\neq i}^{N} \frac{x - x_j}{x_i - x_j}, \quad i = 0, \dots, N.$$

$$(1.29)$$

Note that for $N > 1$, P_χ is not a positive operator.
It is easy to see that

$$\left| f(x) - \sum_{i=0}^{N} l_i(x)f(x_i) \right| \leq \left(1 + \sum_{i=0}^{N} |l_i(x)| \right) E_N(f), \quad (1.30)$$

with

$$E_N(f) = \inf_{p \in \mathcal{P}_N} \max_{x \in [a,b]} |f(x) - p(x)|$$

the error of the best approximation of f by polynomials from $\mathcal{P}_N$.

It is well known that $P_{\chi_N}(f, x)$ does not necessarily converge to $f(x)$ as $N \to \infty$, since the Lebesgue constant $\sum_{i=0}^{N} |l_i(x)|$ may increase faster than

$E_N(f)$ decreases when $N \to \infty$. Yet for Lipschitz continuous functions there is a special choice of points which yields convergence.

Let the interpolation points χ be the roots of the Chebyshev polynomial of degree $N + 1$ on $[-1, 1]$. For these points

$$\sum_{i=0}^{N} |l_i(x)| \le C \log N, \quad x \in [-1, 1].$$

Since $E_N(f) \le C\omega_{[a,b]}(f, 1/N)$ we obtain from (1.30) for a Lipschitz continuous function f

$$\left| f(x) - \sum_{i=0}^{N} l_i(x) f(x_i) \right| \le \frac{C \log N}{N}, \quad x \in [-1, 1], \tag{1.31}$$

and the error in the interpolation of such a function at the roots of the Chebyshev polynomials tends to zero as $N \to \infty$.

1.3.3. *Spline subdivision schemes*

A spline subdivision scheme of order m, $m \ge 1$, in the scalar setting refines the values

$$\mathbf{f}^{k-1} = \{f_\alpha^{k-1} | \alpha \in \mathbb{Z}\} \subset \mathbb{R},$$

and defines $\mathbf{f}^k$ by

$$f_\alpha^k = \sum_{\beta \in \mathbb{Z}} a_{\alpha - 2\beta}^m f_\beta^{k-1}, \quad \alpha \in \mathbb{Z}, \quad k = 1, 2, 3, \ldots \tag{1.32}$$

with the mask

$$a_\alpha^m = \begin{cases} \binom{m}{\alpha} \Big/ 2^{m-1}, & \alpha = 0, 1, \ldots, m \\ 0, & \alpha \in \mathbb{Z} \setminus \{0, 1, \ldots, m\}. \end{cases} \tag{1.33}$$

By (1.32) and (1.33) f_α^k is an average of two or more values from $\mathbf{f}^{k-1}$. It is well known that the values $\mathbf{f}^k$ can be obtained by the Lane–Riesenfeld algorithm, consisting of one step of second order spline subdivision followed by a sequence of $m - 2$ binary averaging. Thus, we first define

$$f_{2\alpha}^{k,1} = f_{\alpha}^{k-1}, \quad f_{2\alpha+1}^{k,1} = \frac{1}{2}(f_{\alpha}^{k-1} + f_{\alpha+1}^{k-1}), \quad \alpha \in \mathbb{Z}. \qquad (1.34)$$

Then for $2 \le j \le m - 1$ we define the averages

$$f_{\alpha}^{k,j} = \frac{1}{2}(f_{\alpha}^{k,j-1} + f_{\alpha+1}^{k,j-1}), \quad \alpha \in \mathbb{Z}. \qquad (1.35)$$

The final values at level k are

$$f_{\alpha}^{k} = f_{\alpha}^{k,m-1}, \quad \alpha \in \mathbb{Z}. \qquad (1.36)$$

For example, in case $m = 3$, one step of averaging yields the Chaikin algorithm

$$f_{2\alpha}^{k} = \frac{3}{4}f_{\alpha}^{k-1} + \frac{1}{4}f_{\alpha+1}^{k-1},$$

$$f_{2\alpha+1}^{k} = \frac{1}{4}f_{\alpha}^{k-1} + \frac{3}{4}f_{\alpha+1}^{k-1}.$$

At each level k, the piecewise linear function, interpolating the data $(2^{-k}\alpha, f_{\alpha}^{k})$, $\alpha \in \mathbb{Z}$, is defined on $\mathbb{R}$, by

$$f^{k}(t) = (\alpha + 1 - 2^{k}t)f_{\alpha}^{k} + (2^{k}t - \alpha)f_{\alpha+1}^{k}, \quad t \in 2^{-k}(\alpha, \alpha + 1), \qquad (1.37)$$

with $\{f_{\alpha}^{k}\}_{\alpha \in \mathbb{Z}}$ the values generated by the m-th order spline subdivision scheme.

Note that by (1.37), every value of $f^{k}(t)$ is a weighted average of two consecutive elements in $\mathbf{f}^{k}$. The sequence $\{f_{m}^{k}(t)\}_{k \in \mathbb{Z}_{+}}$ converges uniformly to a continuous function denoted by $f_{m}^{\infty}(t)$, which is the limit function of the subdivision scheme. Thus the limit function of any spline subdivision scheme can be described in terms of binary averages only.

Remark 1.3.6

(i) The limit function $f_{m}^{\infty}(t)$ is in C^{m-2}, and can be written in terms of the B-spline, b_m, as

$$f_{m}^{\infty} = \sum_{\alpha \in \mathbb{Z}} f_{\alpha}^{0} b_m(\cdot - \alpha). \qquad (1.38)$$

This, together with (1.25), implies that f_{m}^{∞} at each point is a convex combination of m initial values.

Since by (1.38)

$$\sum_{\alpha \in \mathbb{Z}} b_m(Nx - \alpha) f\left(\frac{\alpha}{N}\right) = f_m^\infty(Nx),$$

with f_m^∞ the limit function starting from $f_\alpha^0 = f\left(\frac{\alpha}{N}\right)$, $\alpha \in \mathbb{Z}$, comparison with (1.24) leads to the conclusion that the Lane–Riesenfeld algorithm provides an additional method to the de Boor algorithm for evaluating Schoenberg spline operators by repeated binary averages.

(ii) The rate of convergence of the sequence $\{f^k(x)\}_{k=1}^\infty$ to $f_m^\infty(x)$ is

$$\left| f^k(x) - f_m^\infty(x) \right| \leq C_m \max_{\alpha \in \mathbb{Z}} \left| f_{\alpha+1}^0 - f_\alpha^0 \right| \frac{1}{2^k}$$

with C_m a constant depending on the scheme only.

(iii) Spline subdivision schemes have the property of preservation of monotonicity, namely for monotonically non-decreasing initial data $f_\alpha^0 \leq f_{\alpha+1}^0$, $\alpha \in \mathbb{Z}$ the limit function f_m^∞ is monotone non-decreasing.

1.4. Bibliographical Notes

Basic notions in Approximation Theory, including moduli of continuity and smoothness for real-valued functions are discussed, e.g., in [31, 75, 85]. On geodesic metric spaces and geodesic lines we refer the reader to [19, 74]. Piecewise constant approximation to a bivariate function, based on its level sets, is studied in [32].

Various approximation operators for univariate real-valued functions are studied, e.g., in [31, 61]. Also, the Korovkin theorem on the convergence of positive operators can be found there. The quantitative Korovkin type theorems for real-valued functions are due to [49]. That λ in (1.17) is $O(\frac{1}{N})$ for the Bernstein operator can be seen in [75]. The existence and non-uniqueness of a representation in terms of repeated weighted binary averages of sample-based operators exact for constants, is proved, e.g., in [92]. Error estimates for the Bernstein operators can be found in Chapter 10 of [31], and the de Casteljau algorithm for their evaluation in Chapter 3 of [80]. A detailed study of spline operators is given in [18]; in particular, error estimates for the Schoenberg operators can be found in Chapter XII. The de Boor algorithm for the evaluation of Schoenberg operators is detailed in Chapter 5 of [80]. Interpolation operators and the error estimates for interpolation at the Chebyshev points can be found in [75]. The rate of the error of best approximation by polynomials is given e.g. in Chapter 7 of [31]. The Lane–Riesenfeld algorithm for spline subdivision schemes is due to [62]. More about subdivision schemes can be found in [22, 40, 47].

Chapter 2

On Sets

The sets considered throughout the book are compact. The purpose of this chapter is to provide preliminary information on sets, operations between sets and parametrizations of sets. First we present some notation and definitions on compact sets in $\mathbb{R}^n$ and consider three set operations — the Minkowski sum, the metric average and the metric linear combination — and review some of their properties. Then, the notion of a parametrization of sets and several ways for parametrizing sets are presented. The specific examples of parametrizations given are for convex sets, for star-shaped sets, for sets in $\mathbb{R}$ and for general sets.

2.1. Sets and Operations Between Sets

2.1.1. *Definitions and notation*

We start with notation. We denote by $K(\mathbb{R}^n)$ the collection of all compact non-empty subsets of $\mathbb{R}^n$. All sets considered from now on are from $K(\mathbb{R}^n)$. The set-difference of two sets A and B is $A \setminus B = \{a \in A : a \notin B\}$, the Lebesgue measure of a measurable set A is $\mu(A)$. The Cartesian product of two sets A, B is $A \times B = \{(a, b) : a \in A, b \in B\}$.

By $Co(\mathbb{R}^n)$ we denote the collection of all convex sets in $K(\mathbb{R}^n)$. Recall that a set A is said to be convex if, for any $x, y \in A$ and any $t \in [0, 1]$, the point $tx + (1 - t)y$ is in A. The convex hull of a set A, i.e. the minimal convex set containing A, is denoted by $\mathrm{co}(A)$; the closure of A by $\mathrm{cl}(A)$. The segment between two points $a, b \in \mathbb{R}^n$ is $[a, b] = \mathrm{co}(\{a, b\})$.

Further we use the notation $\langle \cdot, \cdot \rangle$ for the Euclidean inner product, $|\cdot|$ for the Euclidean norm and S^{n-1} for the unit sphere in $\mathbb{R}^n$.

To measure the distance between two sets $A, B \in K(\mathbb{R}^n)$ we use in this book the Hausdorff metric,

$$\text{haus}(A, B) = \max \left\{ \max_{a \in A} \text{dist}(a, B), \ \max_{b \in B} \text{dist}(b, A) \right\},$$

where

$$\text{dist}(a, B) = \min_{b \in B} |a - b|,$$

is the distance from a point $a \in \mathbb{R}^n$ to a set B.

It is well known that the spaces $K(\mathbb{R}^n)$ and $Co(\mathbb{R}^n)$ are complete metric spaces with respect to the Hausdorff metric.

Note that for A and B convex sets in $\mathbb{R}$, $A = [a_1, a_2]$, $B = [b_1, b_2]$,

$$\text{haus}(A, B) = \max\{|a_1 - b_1|, |a_2 - b_2|\}. \tag{2.1}$$

For $A \in K(\mathbb{R}^n)$, we use the notation $\|A\| = \max_{a \in A} |a|$ and denote by ∂A the boundary of A. The **set of projections** of $a \in \mathbb{R}^n$ on a set $B \in K(\mathbb{R}^n)$ is

$$\Pi_B(a) = \{b \in B : |a - b| = \text{dist}(a, B)\}.$$

The **projection** of a set $A \in K(\mathbb{R}^n)$ on a set $B \in K(\mathbb{R}^n)$ is

$$\Pi_B(A) = \bigcup_{a \in A} \Pi_B(a).$$

In the following we present three operations between sets, which are used later in the construction of approximation operators for set-valued functions.

2.1.2. *Minkowski linear combination*

Definition 2.1.1 A Minkowski linear combination of the sets $A_0, \ldots, A_N$ with real coefficients $\lambda_1, \ldots, \lambda_N$ is

$$\sum_{i=1}^{N} \lambda_i A_i = \left\{ \sum_{i=1}^{N} \lambda_i a_i \ : \ a_i \in A_i, \ i = 1, \ldots, N \right\}.$$

In particular,

$$\lambda A = \{\lambda a : a \in A\}, \quad A + B = \{a + b, \ a \in A, \ b \in B\}.$$

The set $A + B$ is called the Minkowski sum of A and B, and the set λA is called the product of A by a scalar λ.

A Minkowski convex combination or a Minkowski average of sets is a Minkowski linear combination with $\{\lambda_i\}_{i=1}^N$ non-negative, summing up to 1.

While $\lambda(A + B) = \lambda A + \lambda B$ for any sets $A, B \in K(\mathbb{R}^n)$ and $\lambda \in \mathbb{R}$, the second distributive low $(\lambda + \mu)A = \lambda A + \mu A$ holds only for $A \in Co(\mathbb{R}^n)$ and $\lambda, \mu \geq 0$.

For the convex hull the following identity holds

$$\mathrm{co}(\lambda A + \mu B) = \lambda \mathrm{co} A + \mu \mathrm{co} B, \quad \lambda, \mu \in \mathbb{R}. \tag{2.2}$$

In Chapters 4 and 5 we study approximation operators based on Minkowski convex combinations.

2.1.3. *Metric average*

The following average between two sets was introduced by Artstein.

Definition 2.1.2 For $A, B \in K(\mathbb{R}^n)$ and $t \in [0, 1]$ the t-weighted metric average of A and B is

$$A \oplus_t B = \{ta + (1-t)b : (a,b) \in \Pi(A,B)\}, \quad t \in [0,1],$$

where

$$\Pi(A,B) = \{(a,b) \in A \times B : a \in \Pi_A(b) \text{ or } b \in \Pi_B(a)\}.$$

We call $\Pi(A, B)$ the set of **metric pairs** of $A, B \in K(\mathbb{R}^n)$.

The most important properties of the metric average are listed below. For $0 \leq t \leq 1, 0 \leq s \leq 1$

1. $A \oplus_0 B = B, \quad A \oplus_1 B = A, \quad A \oplus_t B = B \oplus_{1-t} A,$
2. $A \oplus_t A = A,$
3. $A \cap B \subseteq A \oplus_t B \subseteq tA + (1-t)B \subseteq \mathrm{co}(A \cup B),$
4. $A \oplus_t B = tA + (1-t)B, \quad A, B \in Co(\mathbb{R}),$
5. $\mathrm{haus}(A \oplus_t B, A \oplus_s B) = |t - s| \, \mathrm{haus}(A, B).$

In particular, it follows from Properties 1 and 5 that

$$\mathrm{haus}(A \oplus_t B, A) = (1 - t)\mathrm{haus}(A, B),$$
$$\mathrm{haus}(A \oplus_t B, B) = t \, \mathrm{haus}(A, B). \tag{2.3}$$

The fifth property is termed the **metric property**. Note that the analogues of Properties 2 and 5 in the case of Minkowski averages are true only for convex sets, while with the metric average they are valid for general compact sets. The metric average is a geodesic average as defined in (1.10) in the metric space of $X = K(\mathbb{R}^n)$ endowed with the Hausdorff metric. The Minkowski average is a geodesic average but only in $Co(\mathbb{R}^n)$.

In Chapter 6 we investigate approximation operators based on the metric average. The above properties of the metric average lead to the convergence of some of them.

Remark 2.1.3 For $A, B \in Co(\mathbb{R}^n)$

$$\mathrm{co}(A \oplus_t B) \subseteq tA + (1 - t)B, \tag{2.4}$$

with equality for $n = 1$. For $n > 1$, $A \oplus_t B$ is not necessarily convex, and then there is strict inclusion in (2.4).

We give below an example of two sets in $\mathbb{R}^2$ for which strict inclusion in (2.4) holds.

Example 2.1.4 This example is illustrated by Fig. 2.1.5.
In (a) two sets $A = \{(x, 0): x \in [-1, 1]\}$ and $B = \{(0, y): y \in [-1, 1]\}$ are shown.

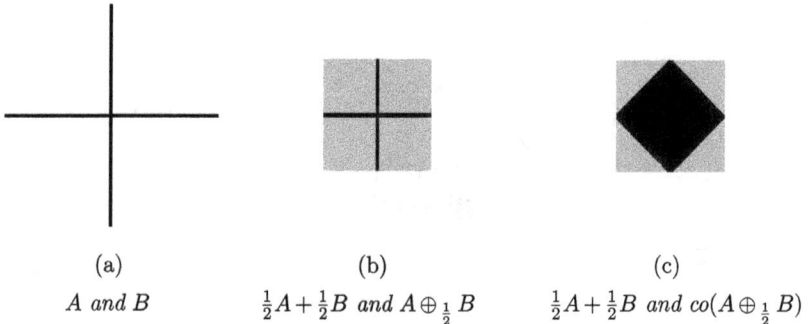

(a)	(b)	(c)
A and B	$\frac{1}{2}A + \frac{1}{2}B$ *and* $A \oplus_{\frac{1}{2}} B$	$\frac{1}{2}A + \frac{1}{2}B$ *and* $\mathrm{co}(A \oplus_{\frac{1}{2}} B)$

Figure 2.1.5   The sets of Example 2.1.4.

Their Minkowski average with weight t is the rectangle

$$tA + (1 - t)B = [-t, t] \times [t - 1, 1 - t].$$

Their metric average is a union of two segments

$$A \oplus_t B = tA \cup (1 - t)B.$$

These two averages for $t = \frac{1}{2}$ are depicted in (b), the first in gray and the second in black. In (c) $\mathrm{co}(A \oplus_{1/2} B)$, in black, is compared with $\frac{1}{2}A + \frac{1}{2}B$ in gray.

Since the metric average is a non-associative binary operation, it cannot straightforwardly be extended to three or more sets. Yet it can be extended to a finite sequence of sets.

2.1.4. *Metric linear combination*

The metric linear combination is an operation on a finite number of ordered compact sets, which extends the metric average. The operation is based on the notion of a metric chain, which generalizes the notion of a metric pair.

Definition 2.1.6 Let $\{A_0, A_1, \ldots, A_N\}$ be a finite sequence of compact sets. A vector $(a_0, a_1, \ldots, a_N)$ with $a_i \in A_i$, $i = 0, \ldots, N$ is called a **metric chain** of $\{A_0, \ldots, A_N\}$ if there exists j, $0 \leq j \leq N$ such that

$$a_{i-1} \in \Pi_{A_{i-1}}(a_i), \ 1 \leq i \leq j \quad \text{and} \quad a_{i+1} \in \Pi_{A_{i+1}}(a_i), \ j \leq i \leq N-1.$$

An illustration of such a metric chain is given in Fig. 2.1.7.

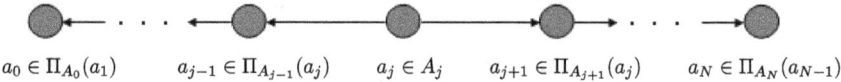

$a_0 \in \Pi_{A_0}(a_1) \qquad a_{j-1} \in \Pi_{A_{j-1}}(a_j) \qquad a_j \in A_j \qquad a_{j+1} \in \Pi_{A_{j+1}}(a_j) \qquad a_N \in \Pi_{A_N}(a_{N-1})$

Figure 2.1.7 A metric chain.

Note that each element of each set A_i, $i = 0, \ldots, N$ generates at least one metric chain. We denote by $CH(A_0, \ldots, A_N)$ the collection of all metric chains of $\{A_0, \ldots, A_N\}$. Note that $CH(A_0, \ldots, A_N)$ depends on the order of the sets. For $N = 1$, $CH(A_0, A_1) = \Pi(A_0, A_1)$.

With this notion of metric chains we can define a set-operation.

Definition 2.1.8 A **metric linear combination** of a finite sequence of sets $A_0, \ldots, A_N$ with coefficients $\lambda_0, \ldots, \lambda_N \in \mathbb{R}$, is

$$\bigoplus_{i=0}^{N} \lambda_i A_i = \left\{ \sum_{i=0}^{N} \lambda_i a_i : (a_0, \ldots, a_N) \in CH(A_0, \ldots, A_N) \right\}. \qquad (2.5)$$

In particular $\bigoplus_{i=0}^{N} 1 \cdot A_i$ is called a **metric sum** and is denoted by $\bigoplus_{i=0}^{N} A_i$.

In the special case $N = 1$ and $\lambda_0, \lambda_1 \in [0, 1]$, $\lambda_0 + \lambda_1 = 1$, the metric linear combination is the metric average. The following are four important properties of the metric linear combination which can be easily derived from the definition:

1. $\displaystyle\bigoplus_{i=0}^{N} \lambda_i A_i = \bigoplus_{i=0}^{N} \lambda_{N-i} A_{N-i},$

2. $\displaystyle\bigoplus_{i=0}^{N} \lambda_i A = \left(\sum_{i=0}^{N} \lambda_i\right) A,$

3. $\displaystyle\bigoplus_{i=0}^{N} \lambda A_i = \lambda\left(\bigoplus_{i=0}^{N} 1 \cdot A_i\right),$

4. $\displaystyle\left(\sum_{i=0}^{N} \lambda_i\right)\left(\bigcap_{i=0}^{N} A_i\right) \subset \bigoplus_{i=0}^{N} \lambda_i A_i \subset \sum_{i=0}^{N} \lambda_i A_i.$

Remark 2.1.9 The metric sum of two sets $A_0 \oplus A_1 = \bigoplus_{i=0}^{1} A_i$ is commutative by the first property, and is not associative in view of (2.5). Similarly one can define the **metric difference** between two sets by $A_0 \ominus A_1 = \bigoplus_{i=0}^{1} \lambda_i A_i$ with $\lambda_0 = 1$, $\lambda_1 = -1$. Then it follows from the second property, that

$$A \ominus A = \{0\}. \tag{2.6}$$

Yet the operation $A \ominus B$ is not the inverse operation of the metric sum as is demonstrated by the following example: $A = [0,1]$, $B = \{0,1\}$.

$$A \ominus B = [-1/2, 1/2], \quad \text{but } (A \ominus B) \oplus B = [-1/2, 1/2] \cup \{3/2\} \neq A.$$

For $\sum_{i=0}^{N} \lambda_i = 1$, from the second property we get

$$\bigoplus_{i=0}^{N} \lambda_i A = A. \tag{2.7}$$

The analogue of this property does not hold for Minkowski linear combinations with some negative coefficients, even for convex sets. This is one reason why only positive operators based on Minkowski sum are considered for set-valued functions with convex images. The metric linear combination allows the adaption of non-positive approximation operators to SVFs. In Chapter 7 we study such operators and also positive ones.

2.2. Parametrizations of Sets

Some classes of sets can be described by a collection of single-valued functions. In this section we bring several examples of such classes, with their corresponding parametrizations.

Definition 2.2.1 Let $\mathcal{A}$ be a collection of sets in $K(\mathbb{R}^n)$. We call a family of functions $\mathcal{G}$ with values in $\mathbb{R}^m$ a **parametrization** of $\mathcal{A}$ if

1. there exists a bijection $T : \mathcal{A} \to \mathcal{G}$,
2. all $g \in \mathcal{G}$ have common domain of definition D.

We denote for every $A \in \mathcal{A}$ its image by $g_A = TA \in \mathcal{G}$.
We term the functions in $\mathcal{G}$ **parametrizing functions** and the elements of D parameters.

The parametrization is called **canonical** if in addition the family $\mathcal{G}$ is closed under convex combinations.

2.2.1. *Induced metrics and operations*

A parametrization $\mathcal{G}$ induces a metric on $\mathcal{A}$. For $A, B \in \mathcal{A}$ this metric is given by

$$d_{\mathcal{G}}(A, B) := \sup_{\xi \in D} |g_A(\xi) - g_B(\xi)| \qquad (2.8)$$

and is termed hereafter **induced metric**.

Moreover, a canonical parametrization also induces a set operation analogous to a convex combination of numbers. For $A, B \in \mathcal{A}$, this operation is defined by

$$tA \uplus (1 - t)B := T^{-1}(tg_A + (1 - t)g_B), \quad t \in [0, 1]$$

and is termed hereafter **induced convex combination**.

This binary operation is easily extended to an induced convex combination of a finite number of sets,

$$\biguplus_{i=1}^{k} t_i A_i = T^{-1} \left(\sum_{i=1}^{k} t_i g_{A_i} \right), \qquad (2.9)$$

for $t_1, \ldots, t_k \in [0, 1]$, satisfying $\sum_{i=1}^{k} t_i = 1$.

Every induced convex combination satisfies the metric property, with respect to the induced metric, namely

$$d_{\mathcal{G}}(tA \uplus (1-t)B, sA \uplus (1-s)B) = |t-s| d_{\mathcal{G}}(A, B).$$

Indeed,

$$d_{\mathcal{G}}(tA \uplus (1-t)B, sA \uplus (1-s)B)$$

$$= \sup_{\xi \in D} |(tg_A(\xi) + (1-t)g_B(\xi)) - (sg_A(\xi) + (1-s)g_B(\xi))|$$

$$= \sup_{\xi \in D} |(t-s)(g_A(\xi) + (s-t)g_B(\xi))| = |t-s| \sup_{\xi \in D} |g_A(\xi) - g_B(\xi)|$$

$$= |t-s| d_{\mathcal{G}}(A, B).$$

Canonical parametrizations facilitate the adaptation of approximation operators defined on real-valued functions to set-valued functions, as is shown in Chapter 4. The rest of Section 2.2 is devoted to examples of parametrizations.

2.2.2. *Convex sets by support functions*

A well-known parametrization of convex compact sets is by their support functions. We recall the definition and some important properties of support functions.

The support function $\delta^*(A, \cdot) : S^{n-1} \to \mathbb{R}$ for $A \in K(\mathbb{R}^n)$ is

$$\delta^*(A, \xi) = \max_{a \in A} \langle \xi, a \rangle, \quad \xi \in S^{n-1}. \tag{2.10}$$

There is a one-to-one correspondence between a convex compact set and its support function. For a non-convex set the support function determines its convex hull.

In this parametrization $\mathcal{G}$ is the set of all support functions of sets in $\mathcal{A} = Co(\mathbb{R}^n)$, and $D = S^{n-1}$.

We list below several important properties of support functions.

For $A, B \in Co(\mathbb{R}^n)$ and $\xi, \tilde{\xi} \in S^{n-1}$,

1. $\delta^*(A + B, \cdot) = \delta^*(A, \cdot) + \delta^*(B, \cdot)$,
2. $\delta^*(\lambda A, \cdot) = \lambda \delta^*(A, \cdot)$, $\lambda \geq 0$,
3. $A \subseteq B \iff \delta^*(A, \cdot) \leq \delta^*(B, \cdot)$,

4. $\text{haus}(A, B) = \max_{\xi \in S^{n-1}} |\delta^*(A, \xi) - \delta^*(B, \xi)|$,

5. $|\delta^*(A, \xi) - \delta^*(A, \tilde{\xi})| \le (\sup_{a \in A} |a|)|\xi - \tilde{\xi}|$.

Remark 2.2.2

(i) There are three important consequences of the above properties of support functions:

— This parametrization is canonical by Properties 1 and 2.
— The induced metric is the Hausdorff metric by Property 4.
— The induced convex combination is the Minkowski convex combination by Properties 1 and 2.

(ii) Definition (2.10) is often extended to $\xi \in \mathbb{R}^n$, and $\delta^*(A, \xi)$ in this case becomes homogenous and sublinear in ξ; namely, it satisfies

$$\delta^*(A, \lambda\xi) = \lambda\delta^*(A, \xi), \ \lambda \ge 0,$$
$$\delta^*(A, \xi + \tilde{\xi}) \le \delta^*(A, \xi) + \delta^*(A, \tilde{\xi}). \tag{2.11}$$

In fact, the set $\mathcal{G}$ is the restriction to S^{n-1} of all functions defined on $\mathbb{R}^n$ which satisfy (2.11).

For general compact sets the existence of a canonical parametrization is an open question, except for compact sets in $\mathbb{R}$ considered in the next section.

2.2.3. *Parametrization of sets in* $\mathbb{R}$

We consider compact sets in $\mathbb{R}$ which are finite unions of compact intervals of positive measure. We denote this class of sets by $\mathcal{A}$. Thus a set $A \in \mathcal{A}$ is given by

$$A = \bigcup_{n=1}^{N} [a_n, b_n], \quad N < \infty, \tag{2.12}$$

where $a_n < b_n$, $n = 1, \ldots, N$, $b_n < a_{n+1}$ for $n = 1, \ldots, N-1$.

For the class $\mathcal{A}_N$ of all sets as in $\mathcal{A}$ with fixed N, there is a parametrization in terms of the boundary points.

For a given set $A \in \mathcal{A}_N$ the corresponding parametrizing function is

$$g_A : \{1, \ldots, N\} \to \mathbb{R}^2, \quad g_A(n) = (a_n, b_n) \in \mathbb{R}^2, \ n \in \{1, \ldots, N\}.$$

It is clear that this parametrization of $\mathcal{A}_N$ is canonical. For the class $\mathcal{A}$ the collection $\{g_A : A \in \mathcal{A}\}$ is not a parametrization, since the domain of g_A depends on the number of segments in A.

Now we introduce a parametrization of $\mathcal{A}$ which is canonical. For $A \in \mathcal{A}$ given by (2.12), the corresponding function $g_A : [0,1] \to \mathbb{R}$ is defined by

$$g_A(\xi) = \min\left\{a \geq a_1 : \frac{\mu([a_1, a] \cap A)}{\mu(A)} = \xi\right\}, \quad \xi \in [0,1], \qquad (2.13)$$

where $\mu(\cdot)$ is the Lebesgue measure of $\mathbb{R}$, or equivalently

$$g_A(\xi) = \begin{cases} a_1, & \xi = 0, \\ a_i + \mu(A)(\xi - \lambda_{i-1}), & \lambda_{i-1} < \xi \leq \lambda_i, \quad i = 1, \ldots, N, \end{cases}$$

$$\lambda_0 = 0, \quad \lambda_i = \frac{1}{\mu(A)} \sum_{j=1}^{i} \mu([a_j, b_j]), \quad i = 1, \ldots, N. \qquad (2.14)$$

From (2.14) it is easy to infer that the parametrization $\mathcal{G}$ consists of all piecewise linear functions defined on $[0,1]$, which are left continuous with a constant non-negative slope. Moreover, (2.14) implies that g_A for $A \in \mathcal{A}$ has a constant slope $\mu(A)$, and discontinuity points $\lambda_1, \ldots, \lambda_{N-1}$. Also, for $g \in \mathcal{G}$, the set $T^{-1}(g)$ is the closure of the image of g.

The next simple example illustrates the relation between A and g_A.

Example 2.2.3 The set A is given by $[a_1, b_1] \cup [a_2, b_2]$ with $a_1 < b_1 < a_2 < b_2$. Let $\lambda = \frac{b_1 - a_1}{b_2 - a_2 + b_1 - a_1}$. Then for $\xi \in [0,1]$

$$g_A(\xi) = \begin{cases} a_1 + \dfrac{b_1 - a_1}{\lambda}\xi, & 0 \leq \xi \leq \lambda, \\[3mm] a_2 + \dfrac{b_2 - a_2}{1 - \lambda}(\xi - \lambda), & \lambda < \xi \leq 1. \end{cases}$$

Note that $\mu(A) = b_2 - a_2 + b_1 - a_1$ and that $\frac{b_1 - a_1}{\lambda} = \frac{b_2 - a_2}{1 - \lambda} = \mu(A)$. The graph of g_A is depicted in Fig. 2.2.4.

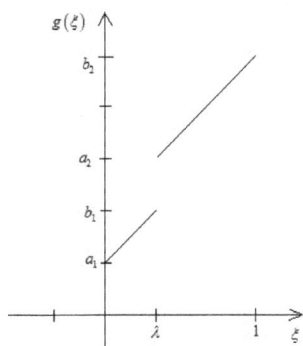

Figure 2.2.4 The graph of g_A in Example 2.2.3.

Since G is closed under convex combinations, the parametrization is canonical. The induced convex combination is a subset of the Minkowski convex combination and, in addition to the metric property, also has the **measure property**, namely

$$\mu(tA \uplus (1-t)B) = t\mu(A) + (1-t)\mu(B). \tag{2.15}$$

The equality in the above formula is easily obtained by considering the slope of $tg_A + (1-t)g_B$.

Next we show that the induced metric bounds the Hausdorff metric.

Proposition 2.2.5 *For $A, B \in \mathcal{A}$*

$$\text{haus}(A, B) \le d_{\mathcal{G}}(A, B) = \sup_{\xi \in [0,1]} |g_A(\xi) - g_B(\xi)|.$$

Proof Since for any $\xi \in [0, 1]$, $g_A(\xi) \in A$, $g_B(\xi) \in B$, we get

$$|g_A(\xi) - g_B(\xi)| \ge \max\{\text{dist}(g_A(\xi), B), \text{dist}(g_B(\xi), A)\}.$$

Since any set $C \in \mathcal{A}$ is the closure of the image of g_C, taking the supremum over $\xi \in [0, 1]$ in the above inequality we get the claim of the proposition. □

It is interesting to note that $tA \uplus (1-t)B = tA + (1-t)B$ for $A, B \in Co(\mathbb{R})$ and that on $Co(\mathbb{R})$ the induced metric coincides with the Hausdorff metric.

2.2.4.   *Star-shaped sets by radial functions*

We recall that a set A is star-shaped if there exists a point $c \in A$ such that for all $a \in A$ the segment $[c, a]$ is contained in A. The point c is called a center of A. The set of all centers of A is called the kernel of A and is a convex set. We denote it by $Ker(A)$.

The mapping of a compact star-shaped set A to the pair $(c_A, \rho_A(\cdot))$, where c_A is the center of mass of $Ker(A)$ and where

$$\rho_A(\xi) = \max\{\beta \in \mathbb{R} : c_A + \beta\xi \in A\}, \ \xi \in S^{n-1},$$

is bijective.

The function $g_A : S^{n-1} \to \mathbb{R}^{n+1}$

$$g_A(\xi) = (c_A, \rho_A(\xi)),$$

defines a parametrization. This parametrization is not canonical in general. Yet in the class of all compact star-shaped sets which are centrally symmetric this parametrization is canonical. Indeed, if $a^* \in A$ is the center of symmetry of A, then it is the center of mass of both A and $Ker(A)$. We note that for $t \in [0, 1]$ the set $tA \uplus (1 - t)B$ is centrally symmetric with center of symmetry $tc_A + (1 - t)c_B$.

The induced metric in this case can be defined as

$$d_{\mathcal{G}}(A, B) = |c_A - c_B| + \sup_{\xi \in S^{n-1}} |\rho_A(\xi) - \rho_B(\xi)|.$$

This metric bounds the Hausdorff metric since for $a = c_A + \rho_A(\xi)\,\xi$ and $b = c_B + \rho_B(\xi)\,\xi$, $|a - b| \leq |c_A - c_b| + |\rho_A(\xi) - \rho_B(\xi)|$.

2.2.5.   *General sets by signed distance functions*

For any compact set $A \in K(\mathbb{R}^n)$ the function

$$\mathrm{sd}(A, \xi) = \begin{cases} \mathrm{dist}(\xi, \partial A), \ \xi \in A \\ -\mathrm{dist}(\xi, \partial A), \ \xi \notin A \end{cases}, \ \xi \in \mathbb{R}^n$$

is called the signed distance function of A.

The set A is easily determined by $A = \{\xi : \mathrm{sd}(A, \xi) \geq 0\}$. Thus the mapping $T(A) = \mathrm{sd}(A, \cdot)$ from $K(\mathbb{R}^n)$ into the set $\mathcal{G}$ of all signed distance functions of sets in $K(\mathbb{R}^n)$ is bijective, and the domain of definition of all functions in $\mathcal{G}$ is $\mathbb{R}^n$, hence $\mathcal{G}$ is a parametrization of $K(\mathbb{R}^n)$.

Remark 2.2.6 The parametrization by signed distance functions is not a canonical parametrization because it is not closed under convex combinations. Yet, one can define a **weak linear combination** with real coefficients

$\lambda_1, \ldots, \lambda_N$ by the set

$$\left\{ x : \sum_{i=1}^{N} \lambda_i sd(A_i, x) \geq 0 \right\}. \tag{2.16}$$

In case $\lambda_i \in [0,1]$, $i = 1, \ldots, N$ and $\sum_{i=1}^{N} \lambda_i = 1$, (2.16) is termed **weak convex combination**.

For sets A, B with a "large" intersection $A \cap B$ and a "small" symmetric difference $(A \setminus B) \cup (B \setminus A)$, this weak convex combination is "geometrically reasonable", which explains the ubiquitous use of this operation in algorithms for reconstruction of objects from their parallel cross-sections. On the other hand, when A and B are disjoint, their average with $t_1 = t_2 = 1/2$ is empty, which is not acceptable.

It is interesting to note that the weak convex combination of two intersecting convex sets is convex, because the sign distance function of a convex set is concave.

The metric induced by the signed distance function bounds the Hausdorff metric from above. Indeed, if we assume that for two compact sets $A \neq B$ the Hausdorff distance is achieved at a point $a \in A$, namely $\text{haus}(A, B) = \text{dist}(a, B)$, then $a \in \partial A \setminus B$ and

$$\text{dist}(a, B) = |\text{sd}(B, a)| = |\text{sd}(B, a) - \text{sd}(A, a)| \leq \|\text{sd}(B, \cdot) - \text{sd}(A, \cdot)\|_\infty.$$

To realize that there is no equality between these two metrics, consider the two sets $A = \{x : |x| \leq 1\}$ and $B = A \setminus \{x : |x| \leq \varepsilon\}$, with $\varepsilon < 1/2$. Indeed $\text{haus}(A, B) = \varepsilon$, while $d_g(A, B) = \text{sd}(A, 0) - \text{sd}(B, 0) = 1 + \varepsilon$.

2.3. Bibliographical Notes

Minkowski linear combinations are well known and used mostly in convex analysis [82, 84] and in mathematical morphology [6, 86].

Definitions and properties of the metric spaces $K(\mathbb{R}^n)$ and $Co(\mathbb{R}^n)$ endowed with the Hausdorff metric can be found in [82–84], in particular the completeness of these two spaces. More about convex sets and convex hulls of sets can also be found in these books.

The metric average is introduced in [5], where its metric property is proved. Some additional properties of this average are studied in [10, 42], and an algorithm for its computation in $\mathbb{R}^1$ is given in the latter. The metric average operation is further extended to metric linear combinations of compact sets in [44]. Algorithms for the computation of the metric average of planar polygons are developed in [55, 66]. The examples calculated there indicate that in many cases the geometry of the metric average is not "in between" the geometry of the two averaged sets, even when the polygons are convex.

The concept of parametrization of sets is closely related to the concept of embedding a collection of sets in a vector space. For a survey of embeddings see Chapter 3 of [6]. In particular, the parametrization of convex compact sets by support functions which was introduced by Hörmander in [54], see also [82, 84], is equivalent to the embedding of $Co(\mathbb{R}^n)$ in the normed vector space of pairs of convex compact sets defined in [81]. This parametrization is used for approximations of set-valued functions and their Aumann integrals in [9, 16, 34, 87]. It is also applied in approximation of control systems and differential inclusions, e.g., in [15, 39, 51, 90, 95].

The parametrizing radial functions of star-shaped sets are studied, e.g., in [89]. Kernels of star-shaped sets are studied in [60].

The signed distance function introduced in [65] is the basis of various methods for reconstruction of objects from cross-sections, see e.g. [27, 52] and references therein.

Chapter 3

On Set-Valued Functions (SVFs)

The notion of set-valued functions, which is central to the book, is presented in this chapter together with several examples. Two important notions related to set-valued functions, selections and representations, are then discussed. In particular two types of representations of $F : [a, b] \to K(\mathbb{R}^n)$ are considered, one based on a parametrization of the sets $F(x)$, $x \in [a, b]$ and one on selections of F. A representation of F allows the introduction of various moduli of continuity and notions of smoothness of F.

In Chapters 4, 8 and 11 we base the adaptation of classical approximation operators to set-valued functions on representations of these functions.

3.1. Definitions and Examples

A set-valued function (SVF, multifunction) F is a mapping with values which are sets. In this book $F : [a, b] \to K(\mathbb{R}^n)$. Recall that $K(\mathbb{R}^n)$, endowed with the Hausdorff metric, is a complete metric space. We investigate the approximation of SVFs using the notions of continuity, bounded variation and Hölder regularity as defined in Section 1.1.

For $F : [a, b] \to K(\mathbb{R}^n)$ we denote by coF the SVF $coF : [a, b] \to Co(\mathbb{R}^n)$ such that $coF(x) = co(F(x))$, and call it the **convex hull of F**. We call the sets $F(x)$, $x \in [a, b]$ **images** of F. The graph of F is

$$Graph(F) = \{(x, y) : y \in F(x), x \in [a, b]\}.$$

It is known that $Graph(F)$ is closed if F is a continuous multifunction with compact images defined on $[a, b]$.

A single-valued function $f : [a, b] \to \mathbb{R}^n$ satisfying $f(x) \in F(x)$, $\forall x \in [a, b]$ is called a selection of F.

A special interesting class of SVFs are multifunctions with images in $Co(\mathbb{R}^n)$. This class has applications in linear control theory and in convex optimization, and has been studied thoroughly.

A simple example of a multifunction with convex images is a segmental function, with each image a segment in $\mathbb{R}$, namely $F : [a, b] \to Co(\mathbb{R})$.

A more sophisticated example of a multifunction with convex images is generated by the solutions of the linear control system

$$\dot{x}(t) = A(t)x + B(t)u(t), \quad x(t_0) = x_0, \ t \geq t_0, \tag{3.1}$$

with $x(t) \in \mathbb{R}^n$, $A(t)$,$B(t)$ matrices of order $n \times n$ and $n \times m$ respectively, and $u(t) \in \mathbb{R}^m$ a piecewise continuous control function, satisfying

$$\forall t \geq t_0, \ u(t) \in U, \quad \text{with } U \in Co(\mathbb{R}^m). \tag{3.2}$$

Denoting the set of all possible solutions $x(\cdot)$ of (3.1) subject to (3.2) by S, the reachable SVF is defined as

$$R(t) = \{x(t) : \ x \in S\}, \quad t \geq t_0.$$

It is not difficult to see that $R(t)$ is a multifunction with convex images. In the non-linear case

$$\dot{x}(t) = f(t, x(t), u(t)), \quad x(t_0) = x_0, \ x(t) \in \mathbb{R}^n, \tag{3.3}$$

with $u(t)$ as above, the images of $R(t)$ are not necessarily convex.

Another example of SVFs with general images is provided by regarding a 3D object M as a univariate SVF

$$F(x) = \{(y, z) \in \mathbb{R}^2 : \ (x, y, z) \in M\}, \ x \in \mathbb{R},$$

namely $F(x_0)$ is the cross-section of M with the plane $x = x_0$ (which can be the empty set).

3.2. Representations of SVFs

In this section we introduce the notion of a representation of multifunctions which belong to a given family $\mathcal{F}$ of SVFs, mapping $[a, b]$ into $K(\mathbb{R}^n)$.

A collection of single-valued functions mapping $[a, b]$ into $\mathbb{R}^m$, for some m,

$$R_\Xi(F) = \{f^\xi(\cdot) : \xi \in \Xi\}, \tag{3.4}$$

with Ξ an index set, is called a **representation** of F if the correspondence between $F \in \mathcal{F}$ and $R_\Xi(F)$ is a bijection. In such a case we denote

$$F \cong R_\Xi(F).$$

Parametrization-based representations

Here we consider representations which are inherently related to parametrizations of the family of sets containing the images of F.

Given a collection of sets $\mathcal{A} \subset K(\mathbb{R}^n)$ and a parametrization $\mathcal{G}$ with a bijection $T : \mathcal{A} \to \mathcal{G}$, the $\mathcal{G}$-**based representation** of a multifunction $F : [a, b] \to \mathcal{A}$ is

$$R^{\mathcal{G}}(F) = \{g_{F(\cdot)}(\xi) : \xi \in D\}, \tag{3.5}$$

with D the domain of definition of the parametrizing functions in $\mathcal{G}$. Note that in this case Ξ in (3.4) is D, and $f^\xi(x) = g_{F(x)}(\xi)$, $x \in [a, b]$. A representation based on a canonical parametrization is called **canonical**.

As an example of a canonical representation we consider the case $\mathcal{A} = Co(\mathbb{R}^n)$ and $\mathcal{G}$ the set of all support functions of sets in $\mathcal{A}$ (see Remark 2.2.2). In this case

$$g_{F(x)}(\xi) = \delta^*(F(x), \xi), \quad \xi \in S^{n-1}. \tag{3.6}$$

It follows from Property 4 of support functions, that a bounded multifunction F with convex images satisfies the equality

$$\omega_{[a,b]}(F, \delta) = \sup_{\xi \in S^{n-1}} \omega_{[a,b]}(\delta^*(F(\cdot), \xi), \delta), \quad \delta > 0, \tag{3.7}$$

with the modulus of continuity of F in the Hausdorff metric (see Section 1.1).

Motivated by (3.7), in Section 3.3 we define moduli of continuity for SVFs in terms of any given parametrization-based representation of F.

Selection-based representations

We call the representation (3.4) of F **selection-based**, if f^ξ for any $\xi \in \Xi$ is a selection of F.

The representation by selections is called Castaing representation if for any $x \in [a, b]$, $F(x) = \text{cl}\{f^\xi(x) : \xi \in \Xi\}$ with Ξ a countable set. An important special case is when

$$F(x) = \{f^\xi(x) : \xi \in \Xi\} \quad \text{for all } x \in [a, b]. \tag{3.8}$$

We call such a representation **complete**.

As an example we consider the collection of sets $\mathcal{A} = K(\mathbb{R}^n)$, and define the selections of $F : [a, b] \to \mathcal{A}$ as

$$f^\xi(x) \in \Pi_{F(x)}(\xi), \quad \xi \in \Xi, \ x \in [a, b], \tag{3.9}$$

with $\Xi = \cup_{z \in [a,b]} F(z) \subset \mathbb{R}^n$.

To see that this is a complete representation, we note that for any $z \in [a, b]$, and for all $\xi \in F(z)$, all the selections $f^\xi(x)$ satisfy $f^\xi(z) = \xi$. In case $\mathcal{A} = Co(\mathbb{R}^n)$, $\Pi_{F(x)}(\xi)$ is a singleton, and there is equality in the definition of $f^\xi(x)$ in (3.9).

In Chapter 8 and in Section 10.3 we investigate two complete representations, and derive approximation results based on these representations. Examples of selection-based representations which are not complete are studied in Sections 10.1 and 10.2.

A representation is both selection-based and parametrization-based, when $g_{F(x)}$ satisfies both $g_{F(x)}(\xi) \in F(x)$ for all $\xi \in D$ and $g_{F(x)} \in \mathcal{G}$ for all $x \in [a, b]$.

For example, such a representation is obtained by generalized Steiner selections of SVFs with convex images. In this example, D is a set of probability measures on the unit ball in $\mathbb{R}^n$. For instance, D may be the set $\mathcal{SM}$ of probability measures with C^1 density, or, alternatively, the set $\mathcal{AM}$ of atomic probability measures concentrated in a single point of the unit sphere in $\mathbb{R}^n$.

To introduce a generalized Steiner selection, we first recall a definition of the Steiner point of a set $A \in \mathbb{R}^n$

$$St(A) = \frac{1}{vol(B_1)} \int_{B_1} y(l, A) dl,$$

where B_1 is the unit ball in $\mathbb{R}^n$, and $y(l, A)$ is a point in the set

$$Y(l, A) = \{a \in A : \langle a, l \rangle = \delta^*(A, l)\}.$$

For a measure $\xi \in D$, the corresponding generalized Steiner point of the set $A \in Co(\mathbb{R}^n)$ is

$$g_A(\xi) = \int_{B_1} St(Y(l, A))\xi(dl).$$

Taking as D a countable dense subset of $\mathcal{SM}$, we get a canonical Castaing representation

$$F(x) = \mathrm{cl}\{g_{F(x)}(\xi) : \xi \in D\}.$$

On the other hand, taking $D = \mathrm{co}(\mathcal{AM})$, we get a complete canonical representation.

It is interesting to note that for SVFs with 1D convex sets as images, the representation based on generalized Steiner selections coincides with the representation based on the canonical parametrization (2.13).

In the next section we define notions of regularity of SVFs in terms of their representations.

3.3. Regularity Based on Representations

Any representation $R_\Xi(F)$ induces notions of regularity and smoothness.

We define a modulus of continuity of F based on (3.4) by

$$\omega^R_{[a,b]}(F, \delta) = \sup_{\xi \in \Xi} \omega_{[a,b]}(f^\xi, \delta), \tag{3.10}$$

and term F to be R-**continuous** whenever $\lim_{\delta \to 0} \omega^R_{[a,b]}(F, \delta) = 0$, which, in view of (3.10), implies that the family $\{f^\xi : \xi \in \Xi\}$ is equicontinuous.

Similarly we define the k-th order ($k \geq 2$) modulus of smoothness of F by

$$\omega^R_{k,[a,b]}(F, \delta) = \sup_{\xi \in \Xi} \omega_{k,[a,b]}(f^\xi, \delta). \tag{3.11}$$

Furthermore, we define F to be R-**smooth of order** k or R-C^k if $f^\xi \in C^k$ for all $\xi \in \Xi$ and if the functions $\left\{\frac{d^k}{dx^k} f^\xi : \xi \in \Xi\right\}$ are equicontinuous. Note that if F is R-C^k then $\omega^R_{k,[a,b]}(F, \delta) = O(\delta^k)$.

In case of a representation based on a parametrization $\mathcal{G}$, we change the notation above replacing R by $\mathcal{G}$. Thus by (3.5)

$$\omega^{\mathcal{G}}_{[a,b]}(F,\delta) = \sup_{\xi \in D} \omega_{[a,b]}(g_{F(\cdot)}(\xi),\delta), \qquad (3.12)$$

and we call $\omega^{\mathcal{G}}_{[a,b]}$ **induced modulus of continuity** of F. In the same way we define $\omega^{\mathcal{G}}_{k,[a,b]}$ and call it **induced modulus of smoothness** of F. A multifunction F is called $\mathcal{G}$-**Hölder**-ν if $\omega^{\mathcal{G}}_{[a,b]}(F,\delta) \leq C\delta^{\nu}$, with $\nu \in (0,1]$. Similarly, F is called $\mathcal{G}$-**smooth of order** k or $\mathcal{G}$-C^k if the functions in

$$R^{\mathcal{G},k}F = \left\{ \frac{d^k}{d(\cdot)^k} f^{\xi} : \xi \in D \right\}$$

are equicontinuous, and it is called $\mathcal{G}$-$C^{k+\nu}$ if the functions in $R^{\mathcal{G},k}F$ are Hölder-ν with the same Hölder constant independent of ξ.

In particular, $\omega^{\mathcal{G}}_{[a,b]}$ in (3.12) can be rewritten as

$$\omega^{\mathcal{G}}_{[a,b]}(F,\delta) = \sup\{d_{\mathcal{G}}\big(F(x_1),F(x_2)\big) : x_1, x_2 \in [a,b], |x_1 - x_2| \leq \delta\}$$

with $d_{\mathcal{G}}$ the induced metric (see Section 2.2.1).

For the example of SVFs with convex images, and the parametrization by support functions, the induced metric is the Hausdorff metric, and (3.10) becomes (3.7).

For the case of complete representations we show in the next lemma that $\omega^{R}_{[a,b]}(F,\delta)$ bounds the modulus of continuity in the Hausdorff metric.

Lemma 3.3.1 *Let* $\{f^{\xi} : \xi \in \Xi\}$ *be a complete representation of a multifunction* $F : [a,b] \to K(\mathbb{R}^n)$. *Then for every* $\delta > 0$

$$\omega_{[a,b]}(F,\delta) \leq \omega^{R}_{[a,b]}(F,\delta). \qquad (3.13)$$

Proof Let $x_1, x_2 \in [a,b]$ and $y \in F(x_1)$ be such that

$$\text{haus}(F(x_1),F(x_2)) = \text{dist}(y,\Pi_{F(x_2)}(y)).$$

Then by (3.8), there exists ξ^* such that $f^{\xi^*}(x_1) = y$. Hence

$$\text{haus}(F(x_1),F(x_2)) = \text{dist}\big(f^{\xi^*}(x_1),\Pi_{F(x_2)}(f^{\xi^*}(x_1))\big)$$
$$\leq |f^{\xi^*}(x_1) - f^{\xi^*}(x_2)|.$$

Finally we get

$$\omega_{[a,b]}(F,\delta) = \sup_{|x_1-x_2|\leq\delta} \mathrm{haus}(F(x_1),F(x_2))$$

$$\leq \sup_{|x_1-x_2|\leq\delta} \sup_{\xi\in\Xi} |f^\xi(x_1) - f^\xi(x_2)|$$

$$= \sup_{\xi\in\Xi} \omega_{[a,b]}(f^\xi,\delta).$$

□

Remark 3.3.2 The result of this lemma holds also for Castaing representations.

It is well known that continuity of a multifunction in the Hausdorff metric does not necessarily imply the existence of continuous selections. Yet, in Chapters 8 and 10 we construct for CBV multifunctions representations by special selections, with modulus of continuity bounded by $C\omega_{[a,b]}(v_F,\delta)$, with C a constant. This guarantees the continuity of the selections in our special representations.

3.4. Bibliographical Notes

For general information on SVFs the reader can consult [8]. Set-valued functions and their selections play a central role in the theory of differential inclusions and their numerical solution (see e.g. [7, 29, 38, 70]), as well as in variational and non-smooth analysis, optimization and modern control theory [26, 59, 71, 72, 83]. An example showing that continuity of a multifunction in the Hausdorff metric does not necessarily imply the existence of continuous selections can be found e.g. in [7], Section 1.6. Representations of SVFs by selections, and in particular Castaing representations, are discussed in [21]. For SVFs with convex images, generalized Steiner selections are introduced in [30] where a Castaing representation based on them is constructed. This construction is developed further and appied for numerical set-valued integration in [14]. Regularity notions with respect to the induced metric are also introduced there, such as Lipschitz continuity and bounded variation.

Various moduli of continuity of multifunctions with values in the metric space $(Co(\mathbb{R}^n), \mathrm{haus}(\cdot,\cdot))$ are defined in [36, 37, 76], and moduli of smoothness of such SVFs are introduced in [34], based on the canonical representation by support functions.

PART II

Approximation of SVFs with Images in $\mathbb{R}^n$

Chapter 4

Methods Based on Canonical Representations

In the previous chapter we introduced the notion of a canonical representation of SVFs. Such a representation is generated by a canonical parametrization $\mathcal{G}$ of a collection of sets $\mathcal{A}$ containing the images of the multifunctions considered. The parametrization $\mathcal{G}$ induces a metric and a convex combination in $\mathcal{A}$. The latter notion allows us to define induced sample-based positive operators for SVFs, while the induced metric facilitates the derivation of bounds on the approximation errors.

This general approach to the approximation of SVFs is motivated by the well-studied case of SVFs with convex images and their representation by support functions. This special case is detailed in Section 4.3, while in Section 4.4 other canonical representations are discussed.

4.1. Induced Operators

Suppose that a canonical representation $R^{\mathcal{G}}$ of a set-valued function F defined on $[a, b]$ with images in $\mathcal{A}$ is given, namely F is determined by its representation

$$R^{\mathcal{G}}(F(\cdot)) = \{f^{\xi}(\cdot) : \xi \in D\}, \tag{4.1}$$

where $f^{\xi}(x) = g_{F(x)}(\xi)$, and D is the domain of definition of the functions in $\mathcal{G}$. Consider a sample-based positive linear operator on real functions of the form

$$(A_\chi f)(x) = \sum_{i=0}^{N} c_i(x) f(x_i), \qquad (4.2)$$

with $\sum_{i=0}^{N} c_i(x) = 1$, $c_i(x) \in [0,1]$, $i = 0, \ldots, N$. The canonical parametrization $\mathcal{G}$ induces on SVFs with images in $\mathcal{A}$ an operator of the form

$$A_\chi^{\mathcal{G}} F(x) = \biguplus_{i=0}^{N} c_i(x) F(x_i), \qquad x \in [a, b], \qquad (4.3)$$

where $\biguplus$ denotes the induced convex combination (2.9).

By the definition of the induced convex combination the set-valued
function $A_\chi^{\mathcal{G}} F$ is represented by the single-valued functions $A_\chi f^\xi(\cdot)$, $\xi \in D$,
namely

$$A_\chi^{\mathcal{G}} F \cong \{(A_\chi f^\xi)(\cdot) : \xi \in D\}. \qquad (4.4)$$

We call the operator (4.3) the **induced operator**.

By definition this operator commutes with the representation, which
can be expressed by the commutativity of the diagram,

$$
\begin{array}{ccc}
F(x) & \longleftrightarrow & \{f^\xi(x) : \xi \in D\} \\
\downarrow A_\chi^{\mathcal{G}} & & \downarrow A_\chi \\
A_\chi^{\mathcal{G}} F(x) & \longleftrightarrow & \{A_\chi f^\xi(x) : \xi \in D\}
\end{array}
$$

For convenience, we use the notation

$$\tilde{F}(x) = A_\chi^{\mathcal{G}} F(x), \quad \tilde{f}^\xi(x) = g_{\tilde{F}(x)}(\xi). \qquad (4.5)$$

In this notation, the above commutativity is expressed by

$$\tilde{f}^\xi(x) = A_\chi f^\xi(x). \qquad (4.6)$$

For example, the induced Bernstein operator of degree N for a given
canonical parametrization $\mathcal{G}$ is

$$B_N^{\mathcal{G}}(F)(x) \cong \left\{ \sum_{i=0}^{N} \binom{N}{i} x^i (1-x)^{N-i} f^\xi \left(\frac{i}{N} \right) : \xi \in D \right\}.$$

4.2. Approximation Results

In this section we obtain approximation results for operators induced by a given canonical parametrization $\mathcal{G}$. Note that this method of adaptation is limited to positive operators A_χ of the form (4.2).

Let $A_\chi^{\mathcal{G}}$ be the induced operator defined in (4.3). In the next proposition we prove that $A_\chi^{\mathcal{G}}F(x)$ approximates SVFs which are continuous in the induced metric, whenever A_χ approximates continuous real-valued functions.

Proposition 4.2.1 *Let the operator A_χ of the form* (4.2) *satisfy for any parametrizing function f^ξ*

$$|A_\chi f^\xi(x) - f^\xi(x)| \le C\omega_{[a,b]}(f^\xi, \phi(x, |\chi|)), \quad x \in [a, b], \qquad (4.7)$$

where $\phi : [a, b] \times \mathbb{R}_+ \to \mathbb{R}_+$ is as in (1.15) *and let $A_\chi^{\mathcal{G}}$ be the corresponding induced operator. Then for a multifunction F with images parametrized by $\mathcal{G}$,*

$$d_{\mathcal{G}}(F(x), A_\chi^{\mathcal{G}}F(x)) \le C\omega_{[a,b]}^{\mathcal{G}}(F, \phi(x, |\chi|)), \quad x \in [a, b],$$

with $\omega_{[a,b]}^{\mathcal{G}}(F, \delta)$ the induced modulus of continuity of F.

Proof Using the definition of the induced metric (2.8), formulas (4.6) and (4.7), and the definition of the induced modulus of continuity (3.12) we get

$$d_{\mathcal{G}}(F(x), \tilde{F}(x)) = \sup_{\xi \in D} |f^\xi(x) - \tilde{f}^\xi(x)| = \sup_{\xi \in D} |f^\xi(x) - A_\chi f^\xi(x)|$$

$$\le C \sup_{\xi \in D} \omega_{[a,b]}(f^\xi, \phi(x, |\chi|)) = C\omega_{[a,b]}^{\mathcal{G}}(F, \phi(x, |\chi|)). \qquad \square$$

Specifying this proposition for the induced Bernstein and Schoenberg operators, using (1.21), (1.26) instead of (4.7), we get the following corollary.

Corollary 4.2.2 *Let $B_N^{\mathcal{G}}$ and $S_{m,N}^{\mathcal{G}}$ be the induced Bernstein and Schoenberg operators respectively. Then*

$$d_{\mathcal{G}}(F(x), B_N^{\mathcal{G}}F(x)) \le C\omega_{[0,1]}^{\mathcal{G}}(F, \sqrt{x(1-x)/N}), \quad x \in [0, 1]. \qquad (4.8)$$

$$d_{\mathcal{G}}(S_{m,N}^{\mathcal{G}}F(x), F(x)) \le \left\lfloor \frac{m+1}{2} \right\rfloor \omega_{[0,1]}^{\mathcal{G}}\left(F, \frac{1}{N}\right), \quad x \in \left[\frac{m-1}{N}, 1\right]. \qquad (4.9)$$

Note that if F is $\mathcal{G}$-Hölder-ν, the above bounds imply rates of convergence $O(N^{-\frac{k}{2}})$ and $O(N^{-\nu})$ for the induced Bernstein operator and the induced Schoenberg operator respectively.

The method of proof of Proposition 4.2.1 can be applied to improve the estimates in this proposition for $\mathcal{G}$-smooth multifunctions, using the induced moduli of smoothness.

Proposition 4.2.3 *Let the operator A_χ of the form (4.2) satisfy for $f \in C[a,b]$*

$$|A_\chi f(x) - f(x)| \le C\omega_{k,[a,b]}(f, \phi(x, |\chi|)). \tag{4.10}$$

Then

$$d_{\mathcal{G}}(F(x), A_\chi^{\mathcal{G}} F(x)) \le C\omega_{k,[a,b]}^{\mathcal{G}}(F, \phi(x, |\chi|)).$$

Note that for a $\mathcal{G}$-$C^{k+\nu}$ multifunction F the error is $O\left(\phi(x, |\chi|^{k+\nu})\right)$.

Korovkin theorem and improved error estimates

The methodology of canonical representations allows us to extend the quantitative Korovkin theorem (Theorem 1.3.2) to SVFs.

Theorem 4.2.4 *Let the operator A_χ of the form (4.2) satisfy*

$$\max_{i=0,1,2} \|A_\chi e_i - e_i\|_\infty \le \lambda, \quad e_i(x) = x^i, \ i = 0, 1, 2. \tag{4.11}$$

Then for any $\mathcal{G}$-continuous multifunction $F \cong \{f^\xi : \xi \in D\}$

$$d_{\mathcal{G}}(F(x), A_\chi^{\mathcal{G}} F(x)) \le C\left(\lambda \|F\|_\infty^{\mathcal{G}} + \omega_{2,[a,b]}^{\mathcal{G}}(F, \sqrt{\lambda})\right), \tag{4.12}$$

with $\|F\|_\infty^{\mathcal{G}} = \sup_{\xi \in D} \|f^\xi\|_\infty$, and C a constant independent of F and x.

Proof In the notation (4.5), we apply Theorem 1.3.2 to the representing family of functions and get

$$|f^\xi(x) - \tilde{f}^\xi(x)| \le C\left(\lambda \|f^\xi\|_\infty + \omega_{2,[a,b]}(f^\xi, \sqrt{\lambda})\right), \quad \xi \in D. \tag{4.13}$$

The constants C and λ in this inequality are independent of x and of the functions $f^\xi(x)$. Taking supremum over $\xi \in D$ in both sides of (4.13), we get the error estimate (4.12), in view of the definitions of the induced metric and the induced modulus of smoothness of second order. $\square$

Thus for F a $\mathcal{G}$-Hölder-ν multifunction, $d_{\mathcal{G}}(F(x), A_\chi^{\mathcal{G}} F(x)) = O(\lambda^{\frac{\nu}{2}})$, while if F is $\mathcal{G}$-$C^{1+\nu}$, the error is $O(\lambda^{\frac{\nu+1}{2}})$.

Specifying Theorem 4.2.4 to the induced Bernstein operators, we get an improvement on the rate of approximation, in view of (1.23) in Remark 1.3.4.

Corollary 4.2.5 *Let* $B_N^{\mathcal{G}}$ *be the induced Bernstein operator. If* F *is* $\mathcal{G}$-$C^{1+\nu}$, *then*

$$d_{\mathcal{G}}(B_N^{\mathcal{G}} F(x), F(x)) = O(N^{-\frac{\nu+1}{2}}).$$

A similar improvement for the symmetric Schoenberg operators follows from Remark 1.3.5.

Corollary 4.2.6 *Let* $\widehat{S}_{m,N}^{\mathcal{G}}$ *be the induced symmetric Schoenberg operator. If* F *is* $\mathcal{G}$-$C^{1+\nu}$, *then*

$$d_{\mathcal{G}}(\widehat{S}_{m,N}^{\mathcal{G}} F(x), F(x)) = O(N^{-1-\nu}).$$

To conclude, the adaptation of approximation operators of type (4.2) based on canonical representations allows us to extend known error estimates for real-valued functions to SVFs, with the error measured in the induced metric.

4.3. Application to SVFs with Convex Images

Here we apply the results of the previous section to the family of SVFs with convex compact images in $\mathbb{R}^n$, using the representation based on support functions. Recall that by Remark 2.2.2 this representation is canonical, the induced convex combinations are the Minkowski ones and the induced metric is the Hausdorff one. Thus, the induced operator (4.3) can be written in terms of a Minkowski convex combination,

$$A_\chi^{Mink} F(x) = \sum_{i=0}^{N} c_i(x) F(x_i). \tag{4.14}$$

Here and in this section $F : [a, b] \to Co(\mathbb{R}^n)$.

The requirements $c_i(x) \geq 0$, $i = 0, \ldots, N$, $\sum_{i=0}^{N} c_i(x) = 1$, for $x \in [a, b]$, ensure the commutativity (4.6) of the operator and the representation by support functions, which is expressed by

$$\delta^*(A_\chi^{Mink} F(x), \xi) = A_\chi \delta^*(F(x), \xi), \quad \xi \in S^{n-1}. \tag{4.15}$$

We note that by (3.7) the induced modulus of continuity is the modulus of continuity in the Hausdorff metric as defined in Section 1.1, and we get the following consequence of Proposition 4.2.1.

Corollary 4.3.1 *Let the operator A_χ of the form (4.2) satisfy (4.7). Then*

$$\text{haus}(F(x), A_\chi^{Mink} F(x)) \leq C\omega_{[a,b]}(F, \phi(x,\delta)).$$

We specialize this result to the Bernstein and Schoenberg operators, with the help of Corollary 4.2.2, replacing there the induced metric by the Hausdorff metric and the induced modulus of continuity by the modulus (1.1) with ρ the Hausdorff metric.

Corollary 4.3.2 *Let B_N^{Mink} and $S_{m,N}^{Mink}$ be the induced Bernstein and Schoenberg operators respectively as in (4.14), namely*

$$B_N^{Mink} F(x) = \sum_{i=0}^{N} \binom{N}{i} x^i (1-x)^{N-i} F\left(\frac{i}{N}\right), \quad x \in [0,1]. \quad (4.16)$$

$$S_{m,N}^{Mink} F(x) = \sum_{i=0}^{N} F\left(\frac{i}{N}\right) b_m(Nx - i), \quad x \in \left[\frac{m-1}{N}, 1\right]. \quad (4.17)$$

Then for $x \in [0,1]$,

$$\text{haus}\left(F(x), B_N^{Mink} F(x)\right) \leq C\omega_{[0,1]}\left(F, \sqrt{x(1-x)/N}\right), \quad (4.18)$$

and for $x \in \left[\frac{m-1}{N}, 1\right]$,

$$\text{haus}\left(F(x), S_{m,N}^{Mink} F(x)\right) \leq \left\lfloor \frac{m+1}{2} \right\rfloor \omega_{[a,b]}\left(F, \frac{1}{N}\right). \quad (4.19)$$

In particular, for a Hölder-ν multifunction F

$$\text{haus}\left(F(x), B_N^{Mink} F(x)\right) = O(N^{-\frac{\nu}{2}}),$$

$$\text{haus}\left(F(x), S_{m,N}^{Mink} F(x)\right) = O(N^{-\nu}). \quad (4.20)$$

Based on Proposition 4.2.3, it is possible to improve these error estimates for a $\mathcal{G}$-smooth multifunction F, namely when the support functions of $F(\cdot)$ have uniformly smooth derivatives.

Corollary 4.3.3 *Let the operator A_χ of the form* (4.2) *satisfy* (4.10). *Then the corresponding operator* (4.14) *satisfies*

$$\text{haus}\,(F(x), A_\chi^{Mink} F(x)) \le C \sup_{\xi \in S^{n-1}} \omega_{k,[a,b]}(\delta^*(F(\cdot), \xi), \phi(x, |\chi|)).$$

It follows that if the functions $\delta^*(F(\cdot), \xi)$, $\xi \in S^{n-1}$ have Hölder-ν k-th derivatives with the same Hölder constant, then the order of the error is $O(\phi(x, |\chi|)^{k+\nu})$.

Furthermore, we can easily reformulate the quantitative Korovkin Theorem 4.2.4 in the case of SVFs with convex compact images.

Corollary 4.3.4 *Let the operator A_χ of the form* (4.2) *satisfy* (4.11). *Then there is a constant C such that for any continuous multifunction F with convex compact images*

$$\begin{aligned}
\text{haus}(F(x), A_\chi^{Mink} F(x)) &\le C(\lambda \|F\|_\infty \\
&+ \sup_{\xi \in S^{n-1}} \omega_{2,[a,b]}(\delta^*(F(\cdot), \xi), \sqrt{\lambda})),
\end{aligned} \quad (4.21)$$

where $\|F\|_\infty = \max_{x \in [a,b]} \|F(x)\| = \max_{x \in [a,b]} \max\{|y| : y \in F(x)\}$.

Using the above result one can also reformulate Corollaries 4.2.5 and 4.2.6.

Corollary 4.3.5 *If the functions $\delta^*(F(\cdot), \xi)$, $\xi \in S^{n-1}$ have Hölder-ν first derivatives with the same Hölder constant, then the error of the induced Bernstein operator is*

$$\text{haus}(F(x), B_N^{Mink} F(x)) = O(N^{-\frac{\nu+1}{2}}), \quad (4.22)$$

and the error of the induced symmetric Schoenberg spline operator is

$$\text{haus}(F(x), \widehat{S}_{m,N}^{Mink} F(x)) = O(N^{-1-\nu}). \quad (4.23)$$

Remark 4.3.6 A Korovkin-type theorem on the convergence only (without error estimates) of sequences of positive linear operators T_N, $N = 1, 2, \ldots$ applied to SVFs with compact convex images is proved by Vitale and in a different way by Mureşan. The operators they consider are not necessarily of the form (4.2), the commutativity (4.15) is not required and (4.11) is replaced by the following two conditions

(a) $\lim_{N \to \infty} T_N(F^i) = F^i$, for $F^i(x) = x^i B_1$, $i = 1, 2$, with B_1 the unit ball,

(b) $\lim_{N \to \infty} \sup\{\text{haus}(T_N F, F) : F(x) \equiv A, \|A\| = 1\} = 0$.

For the induced operators A_χ^{Mink} defined in (4.14), conditions (a) and (b) imply (4.11).

4.4.   Examples and Conclusions

We apply now the results of Section 4.2 to some examples of canonically represented SVFs, and draw conclusions from these examples.

Example 4.4.1 Let $F : [a,b] \to Co(\mathbb{R})$ be defined as $F(x) = [f_1(x), f_2(x)]$, with f_1, f_2 continuous real-valued functions satisfying $f_1(x) \le f_2(x)$, $x \in [a,b]$. We call such an SVF a **segment function**. There are two canonical representations of such a multifunction: one is based on the parametrization by support functions and the other is based on the parametrization (2.13).

In the first parametrization $\mathcal{G}_1$, the domain of parameters (directions in $\mathbb{R}$) is $D = \{-1, 1\}$ and for each x there are only two values of the support function corresponding to these two directions. The canonical representation is

$$F(x) \cong \{-f_1(x), f_2(x)\}.$$

Application of the positive linear operator A_χ on this representation of F results in

$$A_\chi^{\mathcal{G}_1} F(x) \cong \{-A_\chi f_1(x), A_\chi f_2(x)\},$$

which is the representation of the segment function $[A_\chi f_1(x), A_\chi f_2(x)]$, since the positivity of the operator ensures that $A_\chi f_1(x) \le A_\chi f_2(x)$, $x \in [a,b]$.

Consider now the parametrization $\mathcal{G}_2$ of sets in $\mathbb{R}$ defined by (2.13). The domain of parameters is $D = [0,1]$, and

$$F(x) \cong \{f^\xi(x) = (1 - \xi)f_1(x) + \xi f_2(x) : \xi \in [0,1]\}.$$

By (4.4),

$$A_\chi^{\mathcal{G}_2} F(x) \cong \{A_\chi f^\xi(x) = (1 - \xi)A_\chi f_1(x) + \xi A_\chi f_2(x) : \xi \in [0,1]\},$$

which is again the representation of the segment function $[A_\chi f_1(x), A_\chi f_2(x)]$.

Thus both representations yield the same induced operators.

In the next example the multifunction F has compact non-convex images in $\mathbb{R}$. In this case we use only the canonical representation based on the parametrization (2.13), (2.14).

Example 4.4.2 Let $F = [f_0(x), f_1(x)] \cup [f_2(x), f_3(x)]$ with continuous functions f_i, $i = 0, 1, 2, 3$, satisfying for $x \in [a, b]$ $f_i(x) < f_{i+1}(x)$, $i = 0, 1, 2$. Using the notation of Example 2.2.3, let

$$\lambda(x) = \frac{f_1(x) - f_0(x)}{f_1(x) - f_0(x) + f_3(x) - f_2(x)} = \frac{f_1(x) - f_0(x)}{\mu(F(x))}. \qquad (4.24)$$

Then

$$f^\xi(x) = \begin{cases} f_0(x) + \mu(F(x))\xi, & 0 \le \xi \le \lambda(x), \\ f_2(x) + \mu(F(x))(\xi - \lambda(x)), & \lambda(x) < \xi \le 1. \end{cases}$$

The multifunction F is continuous in the induced topology when the functions $\{f^\xi(\cdot) : \xi \in [0, 1]\}$ are uniformly continuous. The last happens only if $\lambda(x)$ is a constant function which is usually not the case. Even the simple multifunction

$$F(x) = [-1, 0] \bigcup [x, 1.5], \quad x \in [0, 1],$$

is not continuous in the induced topology. Indeed, $\lambda(x) = \frac{2}{5-2x}$, and for $\xi \in (\frac{2}{5}, \frac{2}{3})$, $f^\xi(x)$ is discontinuous at the point where $\xi = \lambda(x)$, namely at $x = \frac{5}{2} - \frac{1}{\xi}$.

This example demonstrates that the requirement of continuity of F in the induced metric, needed in Proposition 4.2.1 for approximation in this metric, is very restrictive, and does not hold for general SVFs. A standard way to overcome the problem of discontinuity is to use an integral/averaged modulus of continuity of the representing functions $f^\xi(\cdot)$, in order to define an induced integral/averaged modulus of continuity of the SVF, in a similar way to (3.10). This modulus can be then used in the estimates of the approximation error and may provide approximation without continuity of the representing functions.

In addition to the discontinuity problem, this example reveals a geometrical problem. By (2.14) the parametrizing function of the set $F(x)$ for $x \in [a, b]$ is $g_{F(x)}(\xi)$ which has a point of discontinuity at $\xi = \lambda(x)$ with $\lambda(x)$ defined in (4.24). Note that $0 < \lambda(x) < 1$ for $x \in [a, b]$.

Consider an approximation operator A_χ of the form (4.2). Then by (4.3)

$$A_\chi^{\mathcal{G}} F(x) \cong \left\{ \sum_{i=0}^{N} c_i(x) g_{F(x_i)}(\xi) : \xi \in [0,1] \right\}, \quad x \in [a,b].$$

In case $\lambda(x_i) \neq \lambda(x_j)$ for $i \neq j$, $i, j \in \{0, \ldots, N\}$, then for each $x \in [a,b]$ the representing function $\sum_{i=0}^{N} c_i(x) g_{F(x_i)}(\xi)$ has $N+1$ discontinuity points at $\xi = \lambda(x_i)$, $i = 0, \ldots, N$. Thus for $x \in [a,b]$ the set $A_\chi^{\mathcal{G}} F(x)$ consists of $N+2$ disjoint intervals and the graph of $A_\chi^{\mathcal{G}} F$ is the union of $N+2$ segment functions. This is unacceptable from a geometrical point of view since the graph of F is the union of two segment functions. Therefore, the canonical representation (2.13) is not applicable for approximation.

In Chapters 11 and 12 we discuss alternative representations of SVFs with compact images in $\mathbb{R}$, which provide approximation as well as fidelity to the geometry of the graphs of the SVFs.

The last example in this section is a multifunction with centrally symmetric star-shaped images in $\mathbb{R}^2$. As we have seen in Section 2.2.4, a compact centrally symmetric star-shaped set A in $\mathbb{R}^2$ is parametrized canonically by the function $g_A(\xi) = (c_A, \rho_A(\xi))$, where c_A is the center of symmetry of A and $\rho_A(\xi) = \max\{r \in \mathbb{R} : c_A + r\xi \in A\}$, $\xi \in S^1$.

Example 4.4.3 Using the notation $MA = \{Ma : a \in A\}$ for matrix multiplication of a set $A \subset \mathbb{R}^n$ by a matrix M of order $m \times n$, we define

$$F(x) = \{f(x)\} + M(x)A, \quad x \in [a,b], \tag{4.25}$$

where $f : [a,b] \to \mathbb{R}^2$ is a continuous single-valued function, A is a centrally symmetric set in $\mathbb{R}^2$ with center at the origin, and $M(x)$ is a non-singular 2×2 matrix defined for $x \in [a,b]$.

For the continuity of $F(x)$ in the induced metric, it is not sufficient that $f(x)$ and $M(x)$ are continuous functions. Indeed,

$$g_{F(x)}(\xi) = (f(x), \rho_{F(x)}(\xi)), \quad \xi \in S^1, \quad x \in [a,b],$$

where $\rho_{F(x)}(\xi) = \max\{r : f(x) + r\xi \in F(x)\}$, namely

$$\rho_{F(x)}(\xi) = \max\{r : rM^{-1}(x)\xi \in A\} = \rho_A(M^{-1}(x)\xi).$$

Thus F is continuous in the induced metric if $\rho_A(\cdot)$, $f(\cdot)$ and $M^{-1}(\cdot)$ are continuous. Sufficient conditions for F to be $\mathcal{G} - C^k$ are that $\rho_A(\cdot)$, $f(\cdot)$ and $M^{-1}(\cdot)$ are C^k on $[a,b]$.

For $\mathcal{G}$-continuous or $\mathcal{G} - C^k$ multifunctions of the form (4.25) the theorems proved in Section 4.2 hold, and one can construct approximants in the induced metric. Since the induced metric bounds from above the Hausdorff metric, these approximation theorems provide error bounds also in the Hausdorff metric.

4.5. Bibliographical Notes

To the best of our knowledge, the first approximation results for SVFs are obtained by Vitale [91], who proved the convergence of the Bernstein approximants to multifunctions with convex compact images, using support functions. He also derived a Korovkin-type theorem and noticed the convexification phenomenon in the case of SVFs with non-convex images. We study this phenomenon in the next chapter. Error estimates for approximation of SVFs with convex compact images by positive operators, based on moduli of continuity or on moduli of smoothness of SVFs, are obtained in [9, 34, 76–78]. Results on approximations of multifunctions with convex or general compact images are surveyed in [45, 73]. Moduli of continuity of SVFs have been implemented for error esimates of discrete approximations of differential inclusions in [35, 37].

Approximation results based on the non-canonical parametrization by signed distance functions are obtained in [65] by using the "weak linear combination" defined in Section 2.2.5.

Chapter 5

Methods Based on Minkowski Convex Combinations

Positive operators, based on Minkowski convex combinations, are obtained in Section 4.3. These operators are derived by adapting sample-based positive operators to SVFs with convex images, using the canonical representation of such SVFs in terms of support functions. In this adaptation, convex combinations of numbers in operators of the form $A_\chi f(x) = \sum_{i=0}^{N} c_i(x) f(x_i)$ are replaced by Minkowski convex combinations of sets to obtain induced operators of the form $A_\chi^{Mink} F(x) = \sum_{i=0}^{N} c_i(x) F(x_i)$. It is also shown in Section 4.3 that for SVFs with convex images this adaptation yields good approximation results, with rates of approximation similar to those for real-valued functions.

In this chapter we show that a similar adaptation to convex sets of spline subdivision schemes, based on Minkowski convex combinations, yields converging schemes with similar properties to those of the real-valued schemes. In particular these adapted schemes approximate Hölder-ν multifunctions with convex images. The main part of this chapter is devoted to showing that adaptations based on Minkowski convex combinations fail to approximate SVFs with general images. For that, first we present tools by which the non-convexity of a set can be measured and then use these tools to show the convexity of limit sets of approximation processes based on Minkowski convex combinations. In particular, we show that the sequence of Bernstein operators so adapted, with samples taken from a multifunction F with non-convex images, tends to the convex hull of F. We also show that

53

any spline subdivision scheme based on Minkowski convex combinations converges to a convex-valued multifunction from any initial sets.

5.1. Spline Subdivision Schemes for Convex Sets

In this section we consider the adaptation of spline subdivision schemes (see Section 1.3.3) to convex sets based on Minkowski convex combinations. Given initial convex sets $\{F_\alpha^0 : \alpha \in \mathbb{Z}\}$, we generate at refinement levels $k \geq 1$ sequences of convex sets $\mathbf{F}^k = \{F_\alpha^k : \alpha \in \mathbb{Z}\}$ by

$$F_\alpha^{k+1} = \sum_{\beta \in \mathbb{Z}} a_{\alpha-2\beta}^m F_\beta^k, \quad \alpha \in \mathbb{Z}, \ k = 0, 1, 2, \ldots \tag{5.1}$$

where the coefficients of the mask $\{a_\alpha^m : \alpha \in \mathbb{Z}\}$ are given by

$$a_\alpha^m = \begin{cases} 2^{1-m} \dbinom{m}{\alpha}, & \alpha = 0, 1, \ldots, m, \\ 0, & \alpha \in \mathbb{Z} \backslash \{0, 1, \ldots, m\}. \end{cases} \tag{5.2}$$

Indeed

$$\sum_{\alpha \in \mathbb{Z}} a_{2\alpha}^m = \sum_{\alpha \in \mathbb{Z}} a_{2\alpha+1}^m = 1, \tag{5.3}$$

and thus in (5.1) Minkowski convex combinations are applied.

In the following we consider a spline subdivision scheme of a fixed order m. The scheme is termed convergent if the sequence of Minkowski piecewise linear interpolants to the data $(\alpha 2^{-k}, F_\alpha^k) : \alpha \in \mathbb{Z}$ converges uniformly to a continuous SFV, namely if the sequence $\{F^k(x)\}_{k=0}^\infty$ with

$$F^k(x) = (\alpha + 1 - 2^k x)F_\alpha^k + (2^k x - \alpha)F_{\alpha+1}^k, \quad x \in 2^{-k}[\alpha, (\alpha+1)], \tag{5.4}$$

is a uniform Cauchy sequence.

Now, we prove the convergence of the spline subdivision schemes.

It follows from (5.1) and the properties of support functions stated in Section 2.2.2 that

$$\delta^*(F_\alpha^{k+1}, l) = \sum_{\beta \in \mathbb{Z}} a_{\alpha-2\beta}^m \delta^*(F_\beta^k, l), \quad l \in S^{n-1}. \tag{5.5}$$

Furthermore, since for each x, $F^k(x)$ in (5.4) is a Minkowski convex combination of two sets of $\mathbf{F}^k$, we get for $x \in 2^{-k}[\alpha, (\alpha+1)]$

$$\delta^*(F^k(x), l) = (\alpha + 1 - 2^k x)\delta^*(F_\alpha^k, l) + (2^k x - \alpha)\delta^*(F_{\alpha+1}^k, l). \tag{5.6}$$

Thus by (5.5) and (5.6), for each $l \in S^{n-1}$ the support function $\delta^*(F^k(x), l)$ is the piecewise linear interpolant to the data at refinement level k, obtained by the spline subdivision scheme applied to the initial data $\{\delta^*(F_\alpha^0, l) : \alpha \in \mathbb{Z}\}$.

By the convergence of spline subdivision schemes and by (i) of Remark 1.3.6, the limit obtained from the initial data $\{\delta^*(F_\alpha^0, l) : \alpha \in \mathbb{Z}\}$ for a fixed $l \in S^{n-1}$ is a C^{m-2} function, which we denote by $\gamma(x, l)$. This function is given by

$$\gamma(x, l) = \sum_{\alpha \in \mathbb{Z}} \delta^*(F_\alpha^0, l) b_m(x - \alpha), \tag{5.7}$$

with b_m the B-spline of order m with integer knots and support $[0, m]$. Since the family of support functions is closed under convex combinations, we conclude that for a fixed x, $\{\gamma(x, l) : l \in S^{n-1}\}$ is a parametrization of some convex set by support functions. Denoting this set by

$$F^\infty(x) \cong \{\gamma(x, l) : l \in S^{n-1}\} \tag{5.8}$$

we obtain from Property 4 of support functions, stated in Section 2.2.2, that

$$\text{haus}\left(F^\infty(x), F^k(x)\right) = \max_{l \in S^{n-1}} |\gamma(x, l) - \delta^*(F^k(x), l)|. \tag{5.9}$$

Now by (ii) of Remark 1.3.6, for each $l \in S^{n-1}$

$$|\gamma(x, l) - \delta^*(F^k(x), l)| \leq C_m \frac{1}{2^k} \sup_{\alpha \in \mathbb{Z}} |\delta^*(F_{\alpha+1}^0, l) - \delta^*(F_\alpha^0, l)|.$$

In case of initial data $\mathbf{F}^0$ satisfying $\text{haus}\left(F_{\alpha+1}^0, F_\alpha^0\right) \leq M < \infty$ for all $\alpha \in \mathbb{Z}$, we get by (5.9) that for any $l \in S^{n-1}$

$$|\gamma(x, l) - \delta^*(F^k(x), l)| \leq C \frac{1}{2^k}$$

with C a constant which depends on the scheme and on M. Hence we conclude from (5.8) using again (5.9) that

$$\text{haus}\left(F^\infty(x), F^k(x)\right) \leq C \frac{1}{2^k}$$

showing that $\{F^k(x)\}_{k=0}^\infty$ converges uniformly to $F^\infty(x)$.

Moreover we can get an explicit form of $F^\infty(x)$. By (5.7) we get in view of (5.8)

$$F^\infty(x) = \sum_{\alpha \in \mathbb{Z}} F^0_\alpha b_m(x - \alpha). \tag{5.10}$$

In case the initial data is sampled at equidistant points from a Hölder-ν multifunction H with convex images

$$F^0_\alpha = H(\alpha h), \quad \alpha \in \mathbb{Z}, \ 0 < h \le 1,$$

then $F^\infty(x) = \sum_{\alpha \in \mathbb{Z}} H(\alpha h) b_m(x - \alpha)$ approximates $H(hx)$. Indeed by (i) of Remark 1.3.6 and by (4.17), (4.20) we get the rate of approximation

$$\text{haus}\,(H(x), F^\infty(x/h)) = O(h^\nu), \quad x \in \mathbb{R}.$$

As in the scalar case, the adaptation of spline subdivision schemes to convex sets based on Minkowski convex combinations has shape-preserving properties. It preserves monotonicity in the sense of set inclusion, namely for initial convex sets satisfying

$$F^0_\alpha \subseteq F^0_{\alpha+1}, \quad \alpha \in \mathbb{Z}, \tag{5.11}$$

the limit $F^\infty(x)$ satisfies

$$F^\infty(x) \subseteq F^\infty(y), \quad x < y. \tag{5.12}$$

This property is a direct consequence of the following property of support functions

$$\delta^*(A, l) \le \delta^*(B, l) \quad \text{for all } l \in S^{n-1} \iff A \subseteq B$$

and of (iii) of Remark 1.3.6.

Similarly one can show that for initial sets $\mathbf{F}^0$ satisfying $F^0_{\alpha-1} + F^0_{\alpha+1} \subseteq 2F^0_\alpha$, $\alpha \in \mathbb{Z}$, F^∞ satisfies for all $x, \Delta \in \mathbb{R}$

$$F^\infty(x - \Delta) + F^\infty(x + \Delta) \subseteq 2F^\infty(x).$$

In the rest of this chapter we show that two types of adaptations to SVFs based on Minkowski convex combinations result in convexifying processes. First, we introduce a measure of non-convexity of sets.

5.2. Non-Convexity Measures of a Compact Set

There are several ways to define non-convexity measure, and here we use one of them, which is also called the inner radius.

Definition 5.2.1 For a set A define

$$\text{rad}(A) = \inf_{x \in \mathbb{R}^n} \sup_{a \in A} |x - a|$$

and A^{fin} the collection of all finite subsets of the set A, namely

$$A^{fin} = \left\{ \{a_i\}_{i=1}^k : a_i \in A, i = 1, \ldots, k, \, k \in \mathbb{N} \right\}.$$

Then the non-convexity measure of A is

$$\rho(A) = \sup_{x \in co(A)} \inf \{ \text{rad}(S) : S \in A^{fin}, \, x \in co(S) \}. \tag{5.13}$$

Remark 5.2.2

(i) For $A \in K(\mathbb{R}^n)$, $\text{rad}(A)$ is the minimal radius of a ball containing A.

(ii) More non-convexity measures are known. All of them are non-negative and equal zero for convex sets.

(iii) The non-convexity measure

$$\tilde{\rho}(A) = \sup_{x \in coA} \inf \left\{ \sqrt{\sum_{i=1}^k \alpha_i |x - a_i|^2} : \right.$$

$$\left. a_i \in A, \, x = \sum_{i=1}^k \alpha_i a_i, \sum_{i=1}^k \alpha_i = 1, \alpha_i \in \mathbb{R}_+ \right\}$$

satisfies

$$\tilde{\rho}(A) = \rho(A). \tag{5.14}$$

(iv) One may show that $\text{haus}(A, co(A)) \leq \tilde{\rho}(A)$. Hence by (5.14)

$$\text{haus}(A, co(A)) \leq \rho(A). \tag{5.15}$$

The next basic properties of the measure $\rho(\cdot)$ are obtained directly from its definition.

Proposition 5.2.3 *Let* $A \in K(\mathbb{R}^n)$, *then*

1. $\rho(\lambda A) = |\lambda| \rho(A)$ *for every* $\lambda \in \mathbb{R}$.
2. $\rho(A + \{b\}) = \rho(A)$ *for every* $b \in \mathbb{R}^n$.
3. $\rho(A) \leq \text{rad}(A)$.

The main tool in our analysis is the following generalization of the Shapley–Folkman–Starr theorem due to Cassels.

Theorem 5.2.4 *Let $A_i \in K(\mathbb{R}^n)$, $i = 1, \ldots, l$. Then*

$$\rho\left(\sum_{i=1}^{l} A_i\right) \leq \sqrt{\sum_{i=1}^{l} \rho^2(A_i)}. \tag{5.16}$$

Moreover, if $l > n$ one can take in the right-hand side only the n largest summands. Then (5.16) becomes

$$\rho\left(\sum_{i=1}^{l} A_i\right) \leq \sqrt{\sum_{i=1}^{n} \rho^2(A_i)}, \tag{5.17}$$

where $\rho(A_j) \geq \rho(A_i)$ for $j \leq n < i$.

The next two inequalities follow directly from (5.16) and (5.17), in view of (5.14) and Property 1 in Proposition 5.2.3.

Corollary 5.2.5 *Let $A_i \in K(\mathbb{R}^n)$, $\lambda_i \in \mathbb{R}$, $i = 1, \ldots, l$. Then*

$$\rho\left(\sum_{i=1}^{l} \lambda_i A_i\right) \leq \sqrt{n} \max_{1 \leq i \leq l} |\lambda_i| \rho(A_i) \tag{5.18}$$

and

$$\rho\left(\sum_{i=1}^{l} \lambda_i A_i\right) \leq \mu \max_{1 \leq i \leq l}\{\rho(A_i)\}, \quad \text{with } \mu = \sqrt{\sum_{i=1}^{l} \lambda_i^2}. \tag{5.19}$$

Note that

$$\sum_{i=1}^{l} \lambda_i = 1, \quad \lambda_i \in (0,1), \quad i = 1, 2, \ldots, l \implies \mu = \sqrt{\sum_{i=1}^{l} \lambda_i^2} < 1. \tag{5.20}$$

By (5.19) and (5.20), the process of repeated Minkowski convex combinations with fixed weights produces a sequence of sets with the measure of non-convexity $\rho(\cdot)$, decreasing at least geometrically. This is applied in Section 5.4 to show the convexification of the sets generated by spline subdivision schemes based on Minkowski convex combinations.

On the other hand, by (5.18), averaging with an increasing number of summands l and with coefficients tending to zero, is a convexification process, if $\rho(\cdot)$ of all the sets in the process is bounded. In the next example we apply (5.18) to the simple averaging process with one set.

Example 5.2.6 Let $A \in K(\mathbb{R}^n)$, and define $A_l = \frac{1}{l} \sum_{i=1}^l A$. The obvious inclusions $A \subseteq A_l \subseteq \text{co}(A)$, imply $\text{co}A_l = \text{co}A$, and one easily gets by (5.15) and (5.18)

$$\text{haus}(A_l, \text{co}A) = \text{haus}(A_l, \text{co}A_l) \le \rho(A_l) \le \sqrt{n}\frac{\rho(A)}{l}. \tag{5.21}$$

Therefore $\lim_{l \to \infty} A_l = \text{co}(A)$.

Based on (5.18), we show in the next section that any sequence of sample-based positive operators with increasing number of summands and with coefficients tending to zero tends to a convex limit, when adapted by Minkowski convex combinations, even when the approximated multifunction has non-convex images.

5.3. Convexification of Sequences of Sample-Based Positive Operators

Given a sequence of operators $\{A_N^{Mink}F\}_{N \in \mathbb{N}}$ of the form

$$A_N^{Mink}F(x) = \sum_{k=0}^N c_{k,N}(x)F(x_{k,N}), \quad x \in [a,b], \ N \in \mathbb{N}, \tag{5.22}$$

with $a \le x_{0,N} \le x_{1,N} \le \cdots \le x_{N,N} \le b$ and with $c_{k,N}(x) \ge 0$, $k = 0, 1, \ldots, N$, $\sum_{k=0}^N c_{k,N}(x) = 1$, $x \in [a,b]$, we consider the convergence of this sequence as $N \to \infty$. We show that under mild conditions on the operators and on F, the limit is $\text{co}F$ in a subinterval $[a,b]$, which excludes the possibility of approximating SVFs with general images with these operators .

First we show

Lemma 5.3.1 Let A_N^{Mink} be of the form (5.22), and let $F : [a,b] \to K(\mathbb{R}^n)$. Then

$$\text{haus}\big(A_N^{Mink}F(x), \text{co}F(x)\big)$$

$$\le \sqrt{n} \max_{0 \le k \le N} c_{k,N}(x) \max_{t \in [a,b]} \rho(F(t)) + \text{haus}\big(A_N^{Mink}\text{co}F(x), \text{co}F(x)\big)$$

for any $N \in \mathbb{N}$.

Proof By (2.2) and (5.22)

$$A_N^{Mink} \mathrm{co} F(x) = \mathrm{co} A_N^{Mink} F(x).$$

Using the triangle inequality, one obtains

$$\mathrm{haus}\big(A_N^{Mink} F(x), \mathrm{co} F(x)\big) \le \mathrm{haus}\big(A_N^{Mink} F(x), \mathrm{co} A_N^{Mink} F(x)\big)$$

$$+ \mathrm{haus}\big(A_N^{Mink} \mathrm{co} F(x), \mathrm{co} F(x)\big). \quad (5.23)$$

Now, in view of (5.15)

$$\mathrm{haus}\big(A_N^{Mink} F(x), \mathrm{co} A_N^{Mink} F(x)\big) \le \rho\big(A_N^{Mink} F(x)\big), \quad (5.24)$$

while by (5.18) and (5.22)

$$\rho\big(A_N^{Mink} F(x)\big) \le \sqrt{n} \max_{0 \le k \le N} c_{k,N}(x) \max_{t \in [0,1]} \rho\big(F(t)\big). \quad (5.25)$$

The proof is completed by substituting (5.24) and (5.25) in (5.23). □

The next theorem is the main result of this section.

Theorem 5.3.2 *Let $\{A_N^{Mink}\}_{N \in \mathbb{N}}$ be a sequence of operators of the form* (5.22), *satisfying*

(i) *for any continuous convex-valued multifunction F*

$$\lim_{N \to \infty} A_N^{Mink} F(x) = F(x), \quad x \in [a,b],$$

(ii) *for any $x \in I \subseteq [a,b]$*

$$\lim_{N \to \infty} \max_{0 \le k \le N} c_{k,N}(x) = 0.$$

Then for a continuous multifunction $F : [a,b] \to K(\mathbb{R}^n)$

$$\lim_{N \to \infty} \mathrm{haus}\big(A_N^{Mink} F(x), \mathrm{co} F(x)\big) = 0, \quad x \in I.$$

Proof By Lemma 5.3.1

$$\mathrm{haus}\big(A_N^{Mink} F(x), \mathrm{co} F(x)\big)$$

$$\le \sqrt{n} \max_{0 \le k \le N} c_{k,N}(x) \max_{t \in [a,b]} \rho(F(t)) + \mathrm{haus}\big(A_N^{Mink} \mathrm{co} F(x), \mathrm{co} F(x)\big).$$

Now, since by the continuity of F $\max_{t\in[a,b]} \mathrm{rad}(F(t)) < \infty$, it follows from Property 3 of Proposition 5.2.3 that $\max_{t\in[a,b]} \rho(F(t)) < \infty$. This together with conditions (i) and (ii) of the theorem completes the proof. $\qquad\square$

Operators $\{A_N^{Mink}\}$ satisfying condition (i) of the theorem are studied in Section 4.3 for convex-valued multifunctions. Their approximation rates are derived there, in particular for the Bernstein and the Schoenberg operators.

Moreover, the sequence of Bernstein operators $\{B_N^{Mink}\}_{N\in\mathbb{N}}$ of the form (4.16) also satisfies condition (ii) of Theorem 5.3.2. Indeed, it follows from the local Central Limit Theorem for the binomial distribution that for $x \in I = (0,1)$

$$\lim_{N\to\infty} \sqrt{Nx(1-x)} \max_{0\leq k\leq N} \binom{N}{k} x^k(1-x)^{N-k} = \frac{1}{\sqrt{2\pi}}.$$

Thus for any $x \in I = (0,1)$ there exists $c(x) \in \mathbb{R}_+$ such that

$$\max_{0\leq k\leq N} \binom{N}{k} x^k(1-x)^{N-k} \leq \frac{c(x)}{\sqrt{N}}.$$

Note that $\lim_{x\to0} c(x) = \lim_{x\to1} c(x) = \infty$.

This theory of convexification is not applicable to the Schoenberg operators of the form (4.17), since in the corresponding Minkowski convex combinations the number of non-zero coefficients among $\{b_m(Nx-i)\}_{i=0}^N$ is independent of N, and condition (ii) of Theorem 5.3.2 does not hold.

5.4. Convexification by Spline Subdivision Schemes

Spline subdivision schemes adapted to convex sets are considered in Section 5.1. Here we show that this adaptation, which is based on Minkowski convex combinations, results in schemes that converge to SVFs with convex images, even when the initial sets are not all convex.

Starting from initial sets $\{F_\alpha^0,\ \alpha \in \mathbb{Z}\}$, the adapted spline subdivision scheme of order $m \geq 1$ consists of a sequence of refinement steps,

$$F_\alpha^{k+1} = \sum_{j\in\mathbb{Z}} a_{\alpha-2j}^m F_j^k, \qquad \alpha \in \mathbb{Z},\ \ k = 0,1,2,\dots \qquad (5.26)$$

with a finitely supported mask as in (5.2).

In view of (5.3) we get for $m \geq 3$

$$\nu_m = \sum_\alpha (a_{2\alpha}^m)^2 < 1, \quad \mu_m = \sum_\alpha (a_{2\alpha+1}^m)^2 < 1. \tag{5.27}$$

To define convergence of the scheme, one introduces at the k-th refinement level the piecewise linear multifunction $F^k(\cdot)$ of the form (5.4).

It is shown in Section 5.1 that for initial convex sets $\{F_\alpha^0 : \alpha \in \mathbb{Z}\}$, there exists a set-valued spline function $F^\infty : \mathbb{R} \to K(\mathbb{R}^n)$ which is the uniform limit of the subdivision scheme, namely

$$\lim_{k \to \infty} \sup_{x \in \mathbb{R}} \mathrm{haus}(F^\infty(x), F^k(x)) = 0, \tag{5.28}$$

and that

$$F^\infty(x) = \sum_{\alpha \in \mathbb{Z}} F_\alpha^0 b_m(x - \alpha) \quad \text{for each } x \in \mathbb{R}, \tag{5.29}$$

with b_m the B-spline of order m with integer knots and support $[0, m]$.

In the next theorem we show that for initial data consisting of general sets (not necessarily convex), $\mathbf{F}^0 = \{F_\alpha^0 : \alpha \in \mathbb{Z}\}$, any spline subdivision scheme (5.26) with $m \geq 3$ converges, and the limit multifunction is identical to the one obtained by the same subdivision scheme applied to initial data consisting of the convex hulls of the sets $\mathbf{F}^0$. We prove this result by using Corollary 5.2.5.

Theorem 5.4.1 *The spline subdivision scheme of order m (5.26), when applied to initial data consisting of compact sets $\mathbf{F}^0$ satisfying $\sup\{\rho(F_\alpha^0) : \alpha \in \mathbb{Z}\} = M < \infty$, converges uniformly in the Hausdorff metric to a spline multifunction with convex images of the form*

$$F^\infty(x) = \sum_{\alpha \in \mathbb{Z}} \left(\mathrm{co}F_\alpha^0\right) b_m(x - \alpha). \tag{5.30}$$

Moreover, the rate of convergence is given by

$$\mathrm{haus}(F^k(x), F^\infty(x)) \leq M(\eta_m)^k + \mathrm{haus}(\mathrm{co}F^k(x), F^\infty(x)), \tag{5.31}$$

where $\eta_m = \max\{\sqrt{\mu_m}, \sqrt{\nu_m}\}$.

Proof By (5.4), for $2^{-k}\alpha \leq x \leq 2^{-k}(\alpha+1)$, $F^k(x)$ is a convex combination of F_α^k and $F_{\alpha+1}^k$, hence by (5.19), $\sup_{x \in R} \rho(F^k(x)) \leq \sup_{\alpha \in \mathbb{Z}} \rho(F_\alpha^k)$. Thus, using (5.19), (5.26) and (5.27), we obtain

$$\sup_{x \in R} \rho(F^k(x)) \le \sup_{\alpha \in \mathbb{Z}} \rho(F_\alpha^k) \le \eta_m \sup_{\alpha \in \mathbb{Z}} \rho(F_\alpha^{k-1})$$

$$\le \cdots \le (\eta_m)^k \sup_{\alpha \in \mathbb{Z}} \rho(F_\alpha^0) = M(\eta_m)^k. \tag{5.32}$$

The above inequality gives the convexification rate in the subdivision process.

To see that the limit multifunction is given by (5.30), we denote by $H^k(x)$ the function as in (5.4), obtained from the initial data $\mathbf{H}^0 = \{\mathrm{co}F_\alpha^0 : \alpha \in \mathbb{Z}\}$, and by H^∞ the corresponding limit. It is proved in Section 5.1 that $H^\infty(x)$ equals $F^\infty(x)$ defined in (5.30). Observe that by (2.2), $H^k(x) = \mathrm{co}F^k(x)$ for each x, and that

$$\mathrm{haus}(F^k(x), F^\infty(x)) \le \mathrm{haus}(F^k(x), H^k(x)) + \mathrm{haus}(H^k(x), F^\infty(x))$$

$$= \mathrm{haus}(F^k(x), \mathrm{co}F^k(x)) + \mathrm{haus}(H^k(x), H^\infty(x)).$$

Using (5.15) and (5.32) we finally get

$$\mathrm{haus}(F^k(x), F^\infty(x)) \le \rho(F^k(x)) + \mathrm{haus}(H^k(x), H^\infty(x))$$

$$\le M(\eta_m)^k + \mathrm{haus}(H^k(x), F^\infty(x)),$$

which implies (5.31). By the convergence of the subdivision scheme for convex sets, and since $|\eta_m| < 1$, the proof of the theorem is completed. □

Remark 5.4.2

(i) Estimates of $\mathrm{haus}(H^k(x), F^\infty(x))$ are obtained in Section 5.1.
(ii) The proof of Theorem 5.4.1 applies also for the general class of converging subdivision schemes with non-negative mask coefficients of finite support.

5.5. Bibliographical Notes

Spline subdivision schemes for convex sets based on Minkowski convex combinations are investigated in [41].

Non-convexity measures are studied in [4, 20, 84, 88, 93]. In particular the equality $\rho(A) = \tilde{\rho}(A)$ is proved in the latter.

The convexification phenomenon for the Bernstein operators is first proved in [91] and later for Bernstein-type operators in [43], where the convexification of spline subdivision schemes applied to compact sets is shown. In [91] it is also observed that the piecewise linear interpolant based on Minkowski averages fails to approximate SVFs with general compact sets as images when the mesh size of the interpolation points tends to zero.

Theorem 5.2.4 is a generalization of the Shapley–Folkman–Starr theorem [4, 88] due to Cassels [20].

The local form of the Central Limit Theorem for the binomial distribution can be found e.g. in Sections 11 and 12 of [50].

Chapter 6

Methods Based
on the Metric Average

The metric average was introduced by Artstein, who defined the piecewise linear approximation based on it and showed its convergence when the sampling step size tends to zero. A detailed analysis of this piecewise linear interpolant is presented in the next chapter, and used there in the study of approximation operators adapted by metric linear combinations.

The adaptation of approximation operators to SVFs, based on the metric average, requires a representation of the operators in terms of repeated binary averages. Such a representation exists for any sample-based linear operator which reproduces constants, but is not unique. This non-uniqueness leads to different set-valued operators, and in general it is not clear what are the appropriate adaptations that provide approximation.

Yet, for the Bernstein polynomial operators, Schoenberg spline operators and spline subdivision schemes there exist stable evaluation algorithms by repeated binary weighted averages, as is shown in Sections 1.3.1 and 1.3.3. The adaptations we present here use these algorithms, replacing binary weighted averages of numbers by metric averages of sets. This method of adaptation is satisfactory in case of the Schoenberg operators and spline subdivision schemes, while for Bernstein operators the approximation results we obtain are rather limited.

For the adaptation of the Schoenberg operators, we use the de Boor algorithm. We prove that with this procedural definition of the Schoenberg operators for SVFs, the approximation error bounds for Hölder-ν multi-functions are $O(N^{-\nu})$.

For the adaptation of spline subdivision schemes we use the Lane–Riesenfeld algorithm. We show that the schemes are convergent, and that their limits obtained from samples of a Hölder-ν SVF approximate it with error bounds of order $N^{-\nu}$. Note that these set-valued spline subdivision schemes provide another adaptation of Schoenberg spline operators to SVFs (see Remark 1.3.6).

For the adaptation of a Bernstein polynomial operator we use the de Casteljau algorithm. We prove an approximation result for the restricted case of a Hölder-ν multifunction, with images in $\mathbb{R}$ all of the same topology. The error in this approximation for large degree N is shown to be bounded by a constant multiple of $N^{-\nu/2}$, similar to the case of Hölder-ν real-valued functions. The convergence in the general case is still an open problem. Note that by using the operation of metric linear combinations we get an approximating adaptation of Bernstein operators for general SVFs (see Section 7.4.1).

6.1. Schoenberg Spline Operators

Here we define in a procedural way the set-valued Schoenberg operator of order m for SVFs. We use an extension of de Boor algorithm (1.27) with the average between two numbers replaced by the metric average of two sets.

Definition 6.1.1 Let $F : [0,1] \to K(\mathbb{R}^n)$, $x \in [\frac{m-1}{N}, 1]$, and let $l \in \{m-1, \ldots, N\}$ satisfy $\frac{l}{N} \leq x < \frac{l+1}{N}$. Denote $u = Nx$. The Schoenberg operator of order m based on the metric average, $S_{m,N}^{MA}F(x)$, is defined by the algorithm

$$F_i^0(u) = F(i/N), \quad i = l - m + 1, \ldots, l,$$
$$\text{For } k = 1, \ldots, m-1$$
$$\text{For } i = l - m + k + 1, \ldots, l$$
$$\lambda_i^k = \frac{i + m - k - u}{m - k}$$
$$F_i^k(u) = F_{i-1}^{k-1}(u) \oplus_{\lambda_i^k} F_i^{k-1}(u)$$
$$S_{m,N}^{MA}F(x) = F_l^{m-1}(u).$$

For fixed x, u as in Definition 6.1.1 we use the shorthand notation $F_i^k = F_i^k(u)$. Next, we prove two lemmas and an approximation theorem.

Lemma 6.1.2 *In the notation of Definition 6.1.1, for a fixed x, let*

$$d^k = \max\left\{\text{haus}(F_{i-1}^k, F_i^k) : i = l - m + k + 2, \ldots, l\right\}. \qquad (6.1)$$

Then

$$d^k \le \frac{m-k-1}{m-1} d^0, \quad k = 1, \dots, m-2.$$

Proof It follows from Definition 6.1.1 that

$$\text{haus}(F_i^k, F_i^{k-1}) = \text{haus}(F_{i-1}^{k-1} \oplus_{\lambda_i^k} F_i^{k-1}, F_i^{k-1})$$

$$= \lambda_i^k \, \text{haus}(F_{i-1}^{k-1}, F_i^{k-1}).$$

Thus for $k = 1, \dots, m-1$

$$\text{haus}(F_i^k, F_i^{k-1}) \le \frac{i+m-k-u}{m-k} d^{k-1}. \tag{6.2}$$

In the same way we obtain

$$\text{haus}(F_i^{k-1}, F_{i+1}^k) = \text{haus}(F_i^{k-1}, F_i^{k-1} \oplus_{\lambda_{i+1}^k} F_{i+1}^{k-1})$$

$$= (1 - \lambda_{i+1}^k)\text{haus}(F_i^{k-1}, F_{i+1}^{k-1}).$$

Therefore

$$\text{haus}(F_i^{k-1}, F_{i+1}^k) \le \frac{u-i-1}{m-k} d^{k-1}. \tag{6.3}$$

By the triangle inequality and using the estimates (6.2) and (6.3) we get:

$$\text{haus}(F_i^k, F_{i+1}^k) \le \text{haus}(F_i^k, F_i^{k-1}) + \text{haus}(F_i^{k-1}, F_{i+1}^k)$$

$$\le \frac{m-k-1}{m-k} d^{k-1}.$$

This leads to

$$d^k \le \frac{m-k-1}{m-k} d^{k-1}. \tag{6.4}$$

Now, using (6.4) recursively, we obtain

$$d^k \le \frac{m-k-1}{m-k} d^{k-1}$$

$$\le \frac{m-k-1}{m-k} \cdot \frac{m-(k-1)-1}{m-(k-1)} \cdots \frac{m-2-1}{m-2} \cdot \frac{m-1-1}{m-1} d^0$$

$$= \frac{m-k-1}{m-1} d^0,$$

which proves the claim of the lemma. $\qquad\square$

Lemma 6.1.3 *Under the conditions of Lemma 6.1.2, for any $u \in [l, l+1)$*

$$\mathrm{haus}(F_l^{m-1}(u), F_l^0) \le d^0 \frac{m}{2}. \tag{6.5}$$

Proof For a fixed u we use again the shorthand notation $F_i^k = F_i^k(u)$. By the triangle inequality, (6.2) and Lemma 6.1.2

$$\mathrm{haus}(F_l^{m-1}, F_l^0)$$

$$\le \sum_{k=1}^{m-1} \mathrm{haus}(F_l^{k-1}, F_l^k) \le \sum_{k=1}^{m-1} \frac{m-k+l-u}{m-k} d^{k-1}$$

$$\le \sum_{k=1}^{m-1} \frac{m-k+l-u}{m-k} \cdot \frac{m-(k-1)-1}{m-1} d^0 = \frac{d^0}{m-1} \sum_{k=1}^{m-1} (m-k+l-u)$$

$$= d^0 \left(m+l-u - \frac{1}{m-1} \sum_{k=1}^{m-1} k \right) = d^0 \left(m+l-u - \frac{m}{2} \right) \le d^0 \frac{m}{2},$$

where in the last inequality we use $l \le u \le l+1$. □

As a consequence of the last lemma, we get the approximation result.

Theorem 6.1.4 *Let the set-valued function $F : [0,1] \to K(\mathbb{R}^n)$ be Hölder-ν with a Hölder constant C and let $F_i^0 = F\left(\frac{i}{N}\right)$, $i = 0, 1, \ldots, N$. Then for any $x \in \left[\frac{m-1}{N}, 1\right]$*

$$\mathrm{haus}(S_{m,N}{}^{MA} F(x), F(x)) \le \left(\frac{m}{2} + 1\right) C N^{-\nu}. \tag{6.6}$$

Proof For $x \in \left[\frac{m-1}{N}, 1\right]$, let $l \in \{m-1, \ldots, N-1\}$ be such that $x \in \left[\frac{l}{N}, \frac{l+1}{N}\right]$. Note that for such x the value $S_{m,N}{}^{MA} F(x)$ depends on F_i^0, $i = l-m+1, \ldots, l$. By the triangle inequality we have

$$\mathrm{haus}\left(S_{m,N}{}^{MA} F(x), F(x)\right) \le \mathrm{haus}\left(S_{m,N}{}^{MA} F(x), F_l^0\right) + \mathrm{haus}\left(F_l^0, F(x)\right).$$

Since by Definition 6.1.1, $S_{m,N}{}^{MA} F(x) = F_l^{m-1}(u)$ with $u = Nx$ we obtain from Lemma 6.1.3

$$\mathrm{haus}\left(S_{m,N}{}^{MA} F(x), F(x)\right) \le d^0 \frac{m}{2} + \mathrm{haus}\left(F_l^0, F(x)\right). \tag{6.7}$$

Now, by the Hölder continuity of F,

$$\mathrm{haus}(F_l^0, F(x)) \le C N^{-\nu}, \quad d^0 \le C N^{-\nu}.$$

This together with (6.7) leads to the claim of the theorem. □

To illustrate this approximation result consider the following example.

Example 6.1.5 We construct Schoenberg approximation operators to the multifunction $F(x)$, depicted in gray in Fig. 6.1.6(b) and defined by

$$F(x) = \left\{ y : \max\left\{ 0, \left(\frac{r}{2}\right)^2 - (x-5)^2 \right\} \le y^2 \le r^2 - (x-5)^2 \right\},$$

$$r = 5, \quad x \in [0, 10]. \quad (6.8)$$

(I) Approximation with $S_{4,N}F$.

To illustrate the quality of the approximation, we show in Fig. 6.1.6(b) 41 cross-sections of $S_{4,100}^{MA}F$ depicted in black on the graph of F. The graph of

$$e_h(x) = \text{haus}((S_{4,N}F)(x), F(x))$$

at $x = 4.25$ as function of $h = 10/N$ is also shown in Fig. 6.1.6(a).

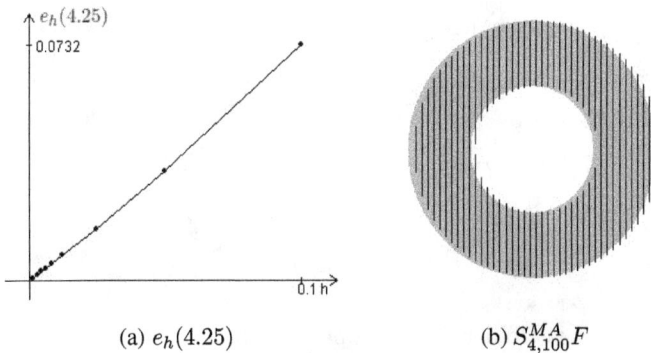

(a) $e_h(4.25)$ (b) $S_{4,100}^{MA}F$

Figure 6.1.6 Approximation of F by $S_{4,N}^{MA}F$.

We note that $e_h(4.25)$ changes almost linearly with h. This is in accordance with Theorem 6.1.4, since, in the neighborhood of $x = 4.25$, F is Lipschitz continuous ($\nu = 1$).

Next, we show in Fig. 6.1.7(a) the graph of the maximal error $\|e_h\|_\infty$ as a function of h. The maximal error is attained at the points $x_1 = 7.5$, $x_2 = 10$, which are depicted in Fig. 6.1.7(b).

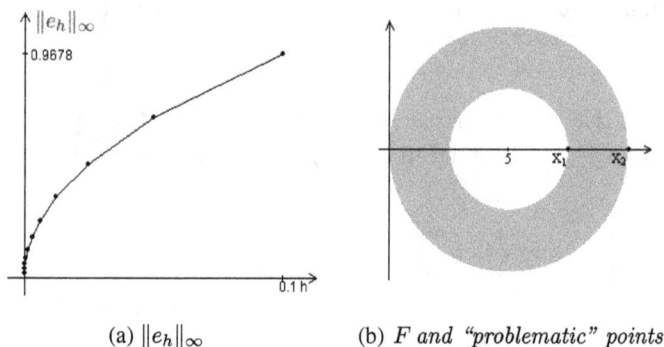

(a) $\|e_h\|_\infty$ (b) *F and "problematic" points*

Figure 6.1.7   Maximal error and "problematic" points.

To verify that the decay of the error in this figure is in accordance with Theorem 6.1.4, we prove that F in (6.8) is Hölder-1/2 in the vicinity of x_1 and x_2. We show this for $x = x_1$.

Consider the inner boundary of $Graph(F)$. From (6.8) it is given by the circle $(x-5)^2 + y^2 = 6.25$, or, equivalently, this boundary is described by the two functions $y_{1,2} = \pm\sqrt{6.25 - (x-5)^2}$. Now, consider the values of F at x_1 and at $x = x_1 - \epsilon$, for some $\epsilon > 0$. From the graph of F it is easy to see that

$$\text{haus}(F(x_1), F(x_1 - \epsilon)) = |y_1(x_1 - \epsilon) - y_1(x_1)| = |y_2(x_1 - \epsilon) - y_2(x_1)|.$$

Thus, it is sufficient to show that y_1 is Hölder -1/2 near x_1. Substituting $x_1 = 7.5$ in the above equality we obtain

$$\text{haus}(F(7.5), F(7.5 - \epsilon)) = |y_1(7.5 - \epsilon) - y_1(7.5)|$$
$$= |\sqrt{6.25 - (7.5 - \epsilon - 5)^2} - 0|$$
$$= \sqrt{5\epsilon - \epsilon^2} \le \sqrt{5\epsilon},$$

and therefore F is Hölder -1/2 in a neighborhood of $x_1 = 7.5$.

Although $Graph(F)$ is symmetric relative to the line $x = 5$, the symmetric points of x_1, x_2 in the graph are not "problematic". This asymmetry follows from the fact that the function $S_{m,N}F$ on the interval $[x_l, x_{l+1}]$ depends on values of F sampled at points left to the interval.

(II) Approximation with $\widehat{S}_{4,N}^{MA}F$.

Here we consider approximation with the symmetric Schoenberg operator $\widehat{S}_{4,N}^{MA}F$.

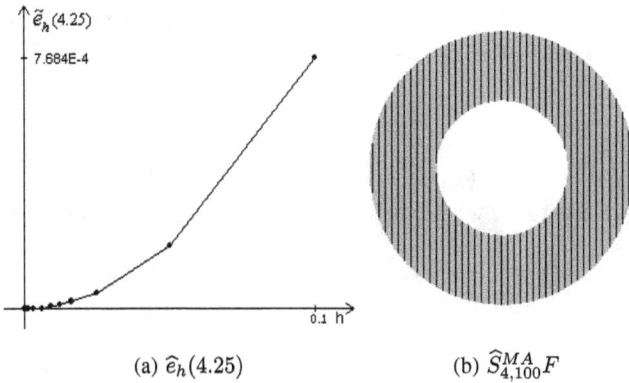

(a) $\widehat{e}_h(4.25)$ (b) $\widehat{S}_{4,100}^{MA}F$

Figure 6.1.8   Approximation of F by $\widehat{S}_{4,N}^{MA}F$.

Figure 6.1.8 is similar to Fig. 6.1.6 but with $\widehat{S}_{4,N}^{MA}F$ replacing $S_{4,N}^{MA}F$. It is easy to observe that the behavior of $\widehat{e}_h(4.25)$ is almost quadratic in h. We conjecture that F is smooth enough at $x = 4.25$ in a sense yet to be defined, and that in general

$$\widehat{e}_h(x) = \text{haus}\big(F(x), \widehat{S}_{2m,N}^{MA}F(x)\big) = O(h^2), \quad m \geq 2 \qquad (6.9)$$

in points of smoothness of F similar to the real-valued case. This improvement over the approximation rate in Theorem 6.1.4 has yet to be proved.

6.2.   Spline Subdivision Schemes

In this section we adapt the Lane–Riesenfeld algorithm (see Section 1.3.3) to general compact sets, replacing averages of numbers by metric averages of sets.

Given a sequence of compact sets in $\mathbb{R}^n$, $\mathbf{F}^0 = \{F_\alpha^0\}_{\alpha \in \mathbb{Z}} \subset K(\mathbb{R}^n)$, we define recursively a sequence of sequences of compact sets,

$$\Big\{\mathbf{F}^k = \big\{F_\alpha^k\big\}_{\alpha \in \mathbb{Z}}\Big\}_{k \in \mathbb{Z}_+}$$

with $\mathbf{F}^k$ the sequence of sets at refinement level k, obtained by the spline subdivision scheme of order m. The sets at each refinement level are obtained by $m - 1$ steps. In the first step the algorithm operates on the sets of the previous refinement level, and in each of the additional $m - 2$ steps it averages the sets generated in the previous step.

First we define the sets at the initial step of refinement level k, from the sets at refinement level $k-1$ in an analogy to (1.34), by

$$F_{2\alpha}^{k,1} = F_{\alpha}^{k-1}, \quad F_{2\alpha+1}^{k,1} = F_{\alpha}^{k-1} \oplus_{\frac{1}{2}} F_{\alpha+1}^{k-1}, \quad \alpha \in \mathbb{Z}. \qquad (6.10)$$

Then in analogy to (1.35) we define recursively the sets in the j-th step by the metric averages with parameter $\frac{1}{2}$ of pairs of consecutive sets in step $j-1$,

$$F_{\alpha}^{k,j} = F_{\alpha}^{k,j-1} \oplus_{\frac{1}{2}} F_{\alpha+1}^{k,j-1}, \quad \alpha \in \mathbb{Z}, \qquad (6.11)$$

for $j = 2, \ldots, m-1$. Note that for $m = 2$, no step of the form (6.11) is performed. Finally the sets at refinement level k are defined similarly to (1.36),

$$F_{\alpha}^{k} = F_{\alpha}^{k,m-1}, \quad \alpha \in \mathbb{Z}. \qquad (6.12)$$

The corresponding piecewise "linear" SVF at refinement level k, interpolating F_{α}^{k} at $\alpha 2^{-k}, \alpha \in \mathbb{Z}$, is defined as

$$F^{k}(t) = F_{\alpha+1}^{k} \oplus_{2^{k}t-\alpha} F_{\alpha}^{k}, \quad \alpha 2^{-k} \le t \le (\alpha+1)2^{-k}, \quad \alpha \in \mathbb{Z}. \qquad (6.13)$$

First we prove two basic metric results stated in two lemmas, which are used in the proofs of the convergence theorem and the approximation result. A major tool in the proofs of the lemmas is the limited form (2.3) of the metric property of the metric average (see Section 2.1.3).

Lemma 6.2.1 *Let*

$$d^{k} = \sup_{\alpha \in \mathbb{Z}} \operatorname{haus}(F_{\alpha}^{k}, F_{\alpha+1}^{k}).$$

Then

$$d^{k} \le d^{0} 2^{-k}, \quad k \in \mathbb{Z}_{+} \qquad (6.14)$$

Proof Denoting by

$$d^{k,j} = \sup_{\alpha \in \mathbb{Z}} \operatorname{haus}(F_{\alpha}^{k,j}, F_{\alpha+1}^{k,j}), \quad j = 1, \ldots, m-1, \qquad (6.15)$$

we obtain from (6.10) and (2.3) that

$$d^{k,1} \le \frac{1}{2} d^{k-1}. \qquad (6.16)$$

Also, by (6.11), the triangle inequality and the metric property, we get for $\alpha \in \mathbb{Z}$,

$$\mathrm{haus}(F_\alpha^{k,j}, F_{\alpha+1}^{k,j}) \leq \mathrm{haus}(F_\alpha^{k,j}, F_{\alpha+1}^{k,j-1}) + \mathrm{haus}(F_{\alpha+1}^{k,j-1}, F_{\alpha+1}^{k,j})$$

$$= \frac{1}{2}\mathrm{haus}(F_\alpha^{k,j-1}, F_{\alpha+1}^{k,j-1}) + \frac{1}{2}\mathrm{haus}(F_{\alpha+1}^{k,j-1}, F_{\alpha+2}^{k,j-1})$$

$$\leq d^{k,j-1}, \quad 2 \leq j \leq m-1,$$

which implies that

$$d^{k,j} \leq d^{k,1}, \quad j = 2, \ldots, m-1. \tag{6.17}$$

This, together with (6.16), yields

$$d^k = d^{k,m-1} \leq \frac{1}{2}d^{k-1}, \tag{6.18}$$

which leads to the claim of the lemma. $\qquad\square$

Lemma 6.2.2 *Let $F^k(\cdot)$ be the piecewise linear interpolant (6.13). Then for $t \in \mathbb{R}$*

$$\mathrm{haus}\left(F^{k+1}(t), F^k(t)\right) \leq Cd^k, \quad k \in \mathbb{Z}_+, \tag{6.19}$$

with $C = \frac{1}{2} + \frac{m}{4}$.

Proof First, we prove the inequality

$$\mathrm{haus}\left(F_{2\alpha}^{k+1}, F_\alpha^k\right) \leq \frac{m-2}{4}d^k, \quad k \in \mathbb{Z}_+. \tag{6.20}$$

By (6.10), (6.12) and the triangle inequality we get

$$\mathrm{haus}(F_\alpha^k, F_{2\alpha}^{k+1}) = \mathrm{haus}(F_{2\alpha}^{k+1,1}, F_{2\alpha}^{k+1,m-1})$$

$$\leq \sum_{j=1}^{m-2} \mathrm{haus}(F_{2\alpha}^{k+1,j}, F_{2\alpha}^{k+1,j+1}).$$

It follows from (6.11) and the metric property that

$$\mathrm{haus}(F_\alpha^k, F_{2\alpha}^{k+1}) \leq \sum_{j=1}^{m-2} \frac{1}{2}\mathrm{haus}(F_{2\alpha}^{k+1,j}, F_{2\alpha+1}^{k+1,j}).$$

Using (6.17) and (6.16), one gets

$$\text{haus}\left(F_\alpha^k, F_{2\alpha}^{k+1}\right) \leq \frac{(m-2)d^{k+1,1}}{2} \leq \frac{(m-2)d^k}{4},$$

which is (6.20).

Now, we prove the claim of the lemma. Let $\alpha 2^{-k} \leq t \leq \left(\alpha + \frac{1}{2}\right) 2^{-k}$. It follows from (6.13) and the metric property that

$$\text{haus}\left(F_\alpha^k, F^k(t)\right) \leq \tfrac{1}{2}d^k, \quad \text{haus}\left(F^{k+1}(t), F_{2\alpha}^{k+1}\right) \leq d^{k+1}. \qquad (6.21)$$

Hence, by (6.20), and (6.21), we obtain

$$\text{haus}(F^{k+1}(t), F^k(t))$$
$$\leq \text{haus}(F^{k+1}(t), F_{2\alpha}^{k+1}) + \text{haus}(F_{2\alpha}^{k+1}, F_\alpha^k) + \text{haus}(F_\alpha^k, F^k(t))$$
$$\leq d^{k+1} + \frac{m-2}{4}d^k + \frac{1}{2}d^k \leq \frac{m+2}{4}d^k.$$

Since for $(\alpha + \frac{1}{2})2^{-k} \leq t \leq (\alpha + 1)2^{-k}$, a similar bound holds (using $F_{2(\alpha+1)}^{k+1}$ instead of $F_{2\alpha}^{k+1}$ and $F_{\alpha+1}^k$ instead of F_α^k in the above derivation), the claim of the lemma follows. $\qquad \square$

We are now ready to prove the two main results.

Theorem 6.2.3 *If the initial sets satisfy*

$$d^0 = \sup_\alpha \text{haus}\left(F_\alpha^0, F_{\alpha+1}^0\right) < \infty,$$

the sequence $\{F^k(\cdot)\}_{k\in\mathbb{Z}_+}$ converges uniformly to a set-valued function $F^\infty(\cdot)$, which is Lipschitz continuous with a Lipschitz constant d_0.

Proof It follows easily from (6.13), the metric property and Lemma 6.2.1, that for any real δ

$$\text{haus}\left(F^k(t+\delta), F^k(t)\right) \leq |\delta|2^k d^k \leq |\delta|d^0. \qquad (6.22)$$

Hence the set-valued functions $\{F^k(\cdot)\}_{k\in\mathbb{Z}_+}$ are uniformly Lipschitz continuous with a Lipschitz constant d^0.

Furthermore, since

$$\text{haus}\left(F^{k+M}(t), F^k(t)\right) \leq \sum_{i=k}^{k+M-1} \text{haus}\left(F^{i+1}(t), F^i(t)\right)$$

we get by Lemmas 6.2.1 and Lemma 6.2.2 that for any positive integer M

$$\text{haus}\left(F^{k+M}(t), F^k(t)\right) \leq C \sum_{i=k}^{k+M-1} d^i \leq C \frac{d^0}{2^{k-1}}, \qquad (6.23)$$

where C is defined in Lemma 6.2.2. This implies that for every t the sequence $\{F^k(t)\}_{k \in \mathbb{Z}_+}$ is a Cauchy sequence in $K(\mathbb{R}^n)$, and since $K(\mathbb{R}^n)$ endowed with the Hausdorff metric is a complete metric space, the sequence $\{F^k(t)\}_{k \in \mathbb{Z}_+}$ tends for each t to a compact set $F^\infty(t)$. The convergence is uniform in t by (6.23). The uniform Lipschitz continuity of $\{F^k(\cdot)\}_{k \in \mathbb{Z}_+}$, which follows from (6.22), yields that $F^\infty(t)$ is also Lipschitz continuous with the same constant d^0. $\qquad \square$

For initial data sampled from a Hölder continuous SVF, we get an approximation result.

Theorem 6.2.4 *Let the set-valued function* $G : \mathbb{R} \to K(\mathbb{R}^n)$ *be Hölder-ν with a Hölder constant L, and let the initial sets be given by $F_\alpha^0 = G(\alpha h)$, $\alpha \in \mathbb{Z}$, then for any $t \in \mathbb{R}$*

$$\text{haus}\left(F^k(t), G(th)\right) \leq C_k h^\nu, \quad k = 0, 1, \ldots \qquad (6.24)$$

where $F^k(\cdot)$ is defined in (6.13) and where

$$C_k = \left(\frac{6+m}{2}\right) L, \quad k \geq 1, \ C_0 = 2L.$$

Proof By the triangle inequality, (6.23), the metric property and the assumption on G, we get for $k \geq 1$ and for t satisfying $\alpha \leq t \leq (\alpha+1)$ that

$$\text{haus}(F^k(t), G(th))$$

$$\leq \text{haus}(F^k(t), F^0(t)) + \text{haus}(F^0(t), F_\alpha^0) + \text{haus}(F_\alpha^0, G(th))$$

$$\leq 2Cd^0 + d^0 + Lh^\nu \leq 2(C+1)Lh^\nu,$$

where C is defined in Lemma 6.2.2, and where we used the Hölder property of G which yields that $d^0 \leq Lh^\nu$. This proves the claim of the theorem, since for $k = 0$ the term with C is missing. $\qquad \square$

Since by (6.24) the sequence $\{F^k\}_{k \geq 1}$ approximates G uniformly in k, we obtain the following direct consequence of Theorem 6.2.4.

Corollary 6.2.5 *Under the notation and assumptions of Theorem 6.2.4*

$$\max_t \text{haus}\left(F^\infty(t), G(th)\right) \leq \frac{6+m}{2} L h^\nu. \qquad (6.25)$$

Remark 6.2.6 Although we do not have the tools to validate it, we expect from numerical simulations that F^∞ is "smoother" than F^0, which approximates G with the same rate but with a smaller constant. Thus for a "smooth" G, the limit F^∞ is advantageous over F^0 as an approximation. Also, we expect F^∞ to give better rates of approximation in this case. The theory is still lacking a notion of smoothness of SVFs which can quantify the above statement on smoothness.

We conclude this section with an example.

Example 6.2.7 A shell between two quarters of spheres is represented by the SVF

$$F(x) = \left\{ (y, z) \in \mathbb{R}^2 \mid z \le 0, \ r(x) \le y^2 + z^2 \le R(x) \right\}, \quad 0 \le x \le 1,$$

where $r(x) = 1 - x^2$, $R(x) = (1.2)^2 - x^2$. From the initial sets (cross-sections) $F(0), F(h), F(2h), \ldots, F(1)$, $h = 0.125$ we approximate this shell by a metric spline subdivision scheme with $m = 2$, corresponding to the Chaikin algorithm.

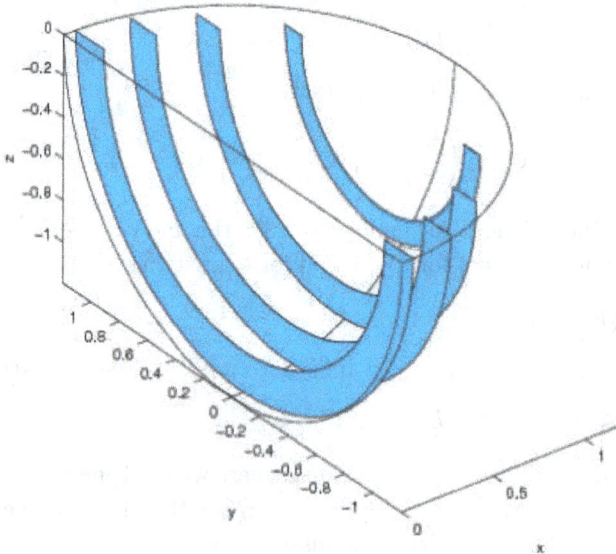

Figure 6.2.8   Cross-sections of approximant.

Figure 6.2.8 displays the cross-sections $F^3 \left(\frac{h}{2} + 0.25i \right)$, $i = 0, 1, 2, 3$ of F^3, obtained after three subdivision iterations from the initial sets. The maximal error between these cross-sections and the corresponding cross-sections of the initial object is 0.0122.

6.3. Bernstein Polynomial Operators

For $F : [0, 1] \to K(\mathbb{R}^n)$, we define procedurally the Bernstein approximation of F in terms of the de Casteljau algorithm (1.20) with the metric average as the basic binary operation.

Definition 6.3.1 For a multifunction F the Bernstein operator of degree N based on the metric average $B_N^{MA} F(x)$ $x \in [0, 1]$, is defined by the algorithm

$$F_i^0(x) = F(i/N), \quad i = 0, \dots, N,$$

For $k = 1, \dots, N$

$$F_i^k(x) = F_i^{k-1}(x) \oplus_{1-x} F_{i+1}^{k-1}(x), \quad i = 0, \dots, N - k$$

$$B_N^{MA} F(x) = F_0^N(x).$$

We do not have a proof of the convergence of $B_N^{MA} F(x)$ to $F(x)$ as $N \to \infty$; also, numerical simulations indicate that if there is convergence, it is extremely slow. Yet, in the special case of SVFs with images in $K(\mathbb{R})$ all of the same topology, we can establish an approximation result, based on the results for B_N^{Mink}. For its proof, we introduce some notions and obtain some auxiliary assertions.

First we prove a preliminary result, which is valid in the general case. For a fixed $x \in [0, 1]$ we use the shorthand notation $F_i^k = F_i^k(x)$.

Lemma 6.3.2 *For a fixed* $x \in [0, 1]$ *let* $F^k = \{F_i^k, i = 0, \dots, N - k\}$ *for* $k = 0, 1, \dots, N - 1$ *be defined as in Definition 6.3.1 and let*

$$d^k = \max_{1 \le i \le N-k} \mathrm{haus}(F_{i-1}^k, F_i^k), \tag{6.26}$$

then

$$d^k \le d^0, \quad k = 1, \dots, N - 1.$$

Proof By the metric property (2.3) of the metric average

$$\mathrm{haus}(F_i^k, F_i^{k-1}) = \mathrm{haus}\left(F_i^{k-1}, F_i^{k-1} \oplus_{1-x} F_{i+1}^{k-1}\right)$$

$$= x \, \mathrm{haus}\left(F_i^{k-1}, F_{i+1}^{k-1}\right) \le x \, d^{k-1}. \tag{6.27}$$

In the same way we obtain

$$\mathrm{haus}\left(F_i^{k-1}, F_{i-1}^k\right) = \mathrm{haus}\left(F_{i-1}^{k-1} \oplus_{1-x} F_i^{k-1}, F_i^{k-1}\right)$$

$$= (1 - x) \, \mathrm{haus}\left(F_{i-1}^{k-1}, F_i^{k-1}\right) \le (1 - x) \, d^{k-1}. \tag{6.28}$$

Now, by the triangle inequality, (6.27) and (6.28) we get

$$\text{haus}\left(F_{i-1}^k, F_i^k\right) \leq \text{haus}\left(F_i^{k-1}, F_{i-1}^k\right) + \text{haus}\left(F_i^{k-1}, F_i^k\right)$$
$$\leq (1-x)d^{k-1} + x\, d^{k-1} = d^{k-1}.$$

Thus

$$d^k \leq d^{k-1},$$

which implies the claim of the lemma. $\square$

Any set $A \in K(\mathbb{R})$ consists of a number of disjoint intervals, some possibly with empty interiors. In the following we assume that the number of intervals, J, is finite. Thus A can be written in the form

$$A = \bigcup_{j=1}^{J} A_j = \bigcup_{j=1}^{J} [\underline{a}_j, \overline{a}_j]$$

with $\underline{a}_1 \leq \overline{a}_1 < \underline{a}_2 \leq \overline{a}_2 < \cdots < \underline{a}_J \leq \overline{a}_J$. We denote this order by $A_1 < \cdots < A_J$.

We introduce a **measure of separation** of such a set,

$$s(A) = \min\left\{\underline{a}_{j+1} - \overline{a}_j : j \in \{1,\ldots,J-1\}\right\} \qquad (6.29)$$

Definition 6.3.3 Two sets $A, B \in K(\mathbb{R})$ are called **topologically equivalent** if each is a union of the same number of disjoint intervals, namely

$$A = \bigcup_{j=1}^{J} A_j, \quad B = \bigcup_{j=1}^{J} B_j, \qquad (6.30)$$

with $A_j, j = 1,\ldots,J$ ($B_j, j = 1,\ldots,J$) disjoint ordered intervals.

Definition 6.3.4 Let $A, B \in K(\mathbb{R})$ be topologically equivalent. The sets A, B are called **metrically equivalent** if

$$\Pi_B(A_j) \subset B_j \quad \text{and} \quad \Pi_A(B_j) \subset A_j, \quad j = 1,\ldots,J. \qquad (6.31)$$

This relation between the two sets is denoted by $A \sim B$.

Note that this relation is reflexive and symmetric, but not transitive.

Remark 6.3.5 The metric average of two metrically equivalent sets A and B is given by

$$A \oplus_t B = \bigcup_{j=1}^{J} (A_j \oplus_t B_j) = \bigcup_{j=1}^{J} (tA_j + (1-t)B_j),$$

where the last equality is a result of Property 4 of the metric average (see Section 2.1.3).

In the next three lemmas, properties of the measure of separation are studied.

Lemma 6.3.6 *If* $A \sim B$, *then for any* $t \in [0,1]$ $A \oplus_t B \sim A$. *Moreover*

$$s(A \oplus_t B) \geq \min(s(A), s(B)). \tag{6.32}$$

Proof The first claim of the lemma is trivial. We prove (6.32). Denote $C = A \oplus_t B$, thus by Remark 6.3.5

$$C = \bigcup_{j=1}^{J} C_j = \bigcup_{j=1}^{J} [\underline{c}_j, \overline{c}_j],$$

where for $j = 1, \ldots, J$, $\underline{c}_j = t\underline{a}_j + (1-t)\underline{b}_j$ and similarly for the value $\overline{c}_j$. Clearly $s(C) = \underline{c}_{j*+1} - \overline{c}_{j*}$, for some $j^* = \{1, \ldots, J-1\}$. Thus by (6.29),

$$s(C) = |\underline{c}_{j^*+1} - \overline{c}_{j^*}| = t|\underline{a}_{j^*+1} - \overline{a}_{j^*}| + (1-t)|\underline{b}_{j^*+1} - \overline{b}_{j^*}|$$

$$\geq t\, s(A) + (1-t)\, s(B) \geq \min(s(A), s(B)),$$

which asserts (6.32). $\qquad\qquad\square$

Lemma 6.3.7 *Let* $A, B \in K(\mathbb{R})$ *be topologically equivalent. If*

$$\mathrm{haus}(A, B) < \frac{\min(s(A), s(B))}{2} \tag{6.33}$$

then A *and* B *are metrically equivalent.*

Proof Assume to the contrary that (6.33) holds, but A, B are not metrically equivalent. First assume that there exists a subset $B_l \in B$ such that two points from B_l have their closest points in A in two subsets of A, say A_j and A_{j+1}. By the continuity of the projection mapping and since B_l is an interval, there exists a point $\tilde{b} \in B_l$ such that $\{\overline{a}_j, \underline{a}_{j+1}\} = \Pi_A(\tilde{b})$.

By the triangle inequality,

$$|\bar{a}_j - \underline{a}_{j+1}| \leq |\tilde{b} - \bar{a}_j| + |\tilde{b} - \underline{a}_{j+1}| = 2\,\mathrm{dist}(\tilde{b}, A). \qquad (6.34)$$

Now, by the definition of the Hausdorff distance, (6.34) and (6.29) we obtain:

$$\mathrm{haus}(A, B) \geq \mathrm{dist}(\tilde{b}, A) \geq \frac{1}{2}s(A)$$

in contradiction to assumption (6.33). Thus for any $l \in \{1, \ldots, J\}$ there exists $j \in \{1, \ldots, J\}$ such that $\Pi_A(B_l) \subseteq A_j$, and by symmetry there exists $k \in \{1, \ldots, J\}$ such that $\Pi_B(A_j) \subseteq B_k$. It remains to prove that $k = l = j$.

Let $a_j \in A_j$ and $b_l \in B_l$ be such that $a_j = \Pi_A(b_l)$. Let $b_k \in B_k$ be such that $b_k = \Pi_B(a_j)$. By the triangle inequality, by the definition of the Hausdorff distance and by (6.33),

$$|b_l - b_k| \leq |b_l - a_j| + |a_j - b_k|$$
$$= \mathrm{dist}(b_l, A) + \mathrm{dist}(a_j, B) \leq 2\mathrm{haus}(A, B) < s(B), \qquad (6.35)$$

implying that $k = l$. By symmetry and since A and B are both of the form (6.30), we conclude that $l = j$. Thus $A \sim B$. $\qquad \square$

Lemma 6.3.8 *Let $F_i^0 \subset K(\mathbb{R})$, $i = 0, 1, \ldots, N$ be topologically equivalent, of the form*

$$F_i^0 = \bigcup_{j=1}^{J} F_{i,j}^0,$$

with $F_{i,j}^0, j = 1, \ldots, J$ disjoint ordered intervals. Define $\{F_i^k\}$ and d^k by Definition 6.3.1 and by (6.26) respectively, and define

$$s^k = \min\left\{s(F_i^k) : i = 0, 1, \ldots, N - k\right\}, \quad k = 0, 1, \ldots, N - 1. \qquad (6.36)$$

If $d^0 < s^0/2$, then

$$d^k < \frac{s^k}{2}, \quad k = 1, \ldots, N - 1, \qquad (6.37)$$

and any pair of sets F_i^k, F_{i+1}^k, $i = 0, \ldots, N - k - 1$, $k = 0, \ldots, N - 1$ is metrically equivalent.

Proof We prove the lemma by induction on k. For $k = 0$ the claim follows from Lemma 6.3.7. We assume that the claim holds for $k - 1$. By

Remark 6.3.5

$$F_i^k = \bigcup_{j=1}^{J} (F_{i,j}^{k-1} \oplus_{1-x} F_{i+1,j}^{k-1}), \quad i = 0, \ldots, N - k. \tag{6.38}$$

Moreover, Lemma 6.3.6 and (6.36) lead to $F_i^k \sim F_i^{k-1}$, $F_i^k \sim F_{i+1}^{k-1}$, $i = 0, \ldots, N - k$ and to

$$s^k \geq s^{k-1}, \tag{6.39}$$

which in view of Lemma 6.3.2 and the induction hypothesis yields

$$d^k \leq d^{k-1} < \frac{s^{k-1}}{2} \leq \frac{s^k}{2}.$$

Finally, we conclude from Lemma 6.3.7 that any pair of sets F_i^k, F_{i+1}^k, $i = 0, \ldots, N - k - 1$ are metrically equivalent, which completes the inductive step. □

Now we can prove,

Theorem 6.3.9 *Let the multifunction $F : [0,1] \to K(\mathbb{R})$ be Hölder-ν, such that for each $x \in [0,1]$, $F(x) = \cup_{j=1}^{J} F_{[j]}(x)$, with $J > 1$, where $\{F_{[j]}(x)\}_{j=1}^{J}$ are disjoint ordered intervals. Then for N large enough*

$$\mathrm{haus}\left(B_N^{MA}F(x), F(x)\right) \leq O(N^{-\frac{\nu}{2}}), \quad x \in [0,1]. \tag{6.40}$$

Proof The case $J = 1$ is omitted, because in view of Remark 6.3.5 it is proved in Section 4.3.

For $J > 1$, by the assumption on F

$$s^* = \inf_{0 \leq x \leq 1} s(F(x)) > 0.$$

Since F is Hölder-ν, there exists N such that $d^0 < s^*/2$, with d^0 defined by (6.26). Obviously $s^* \leq s^0$, where s^0 is defined in (6.36). Thus $d^0 \leq s^0/2$. Now, by Remark 6.3.5 and Lemma 6.3.8, we get

$$B_N^{MA}F(x) = \bigcup_{j=1}^{J} B_N^{Mink}F_{[j]}(x).$$

Therefore

$$\mathrm{haus}\left(F(x), B_N^{MA}F(x)\right) \leq \max_{1 \leq j \leq J} \mathrm{haus}\left(F_{[j]}(x), B_N^{Mink}F_{[j]}(x)\right),$$

and (6.40) follows from Corollary 4.3.2. □

We illustrate Theorem 6.3.9 with the following example.

Example 6.3.10 Let

$$F(x) = \{\{y : 1 \leq y \leq 0.06x^2 + 2\} \bigcup \{y : 0.1x^2 + 2.5 \leq y \leq 13.5\}\},$$

$$x \in [0, 10].$$

Since the Bernstein operator is defined on functions in $C[0, 1]$, we substitute x by $x/10$ in Definition 6.3.1.

In Fig. 6.3.11(a), (b), (c) the SVF is depicted in gray. Forty one cross-sections of $B_{12}^{MA}F$, $B_{13}^{MA}F$ and $B_{30}^{MA}F$ colored by black are shown in (a), (b) and (c), respectively.

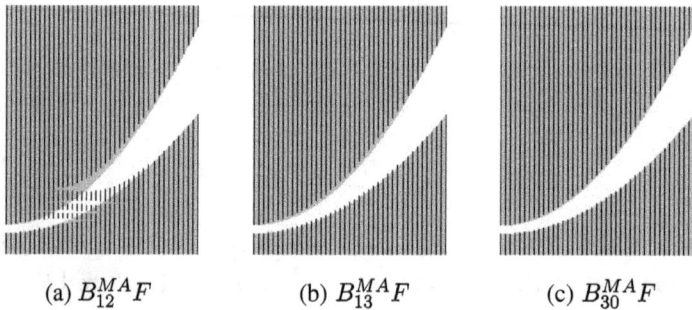

(a) $B_{12}^{MA}F$ (b) $B_{13}^{MA}F$ (c) $B_{30}^{MA}F$

Figure 6.3.11   Comparison between $B_N^{MA}F$, $N = 12, 13, 30$.

Note that the condition $d^0 < s^*/2$, with d^0 defined by (6.26) and s^* as in Theorem 6.3.9, does not hold for $N = 12$, but holds for $N \geq 13$. Figure 6.3.11 shows that for $N = 12$ there is no approximation, while $B_{13}^{MA}F$ is already approximating F. The shape of $B_{12}^{MA}F$ in (a) indicates that some pairs of consecutive sets in $\{F_i^0 : i = 0, \ldots, 12\}$ are not metrically equivalent. It is easily observed that the approximation by $B_{30}^{MA}F$ is better than that by $B_{13}^{MA}F$.

6.4.   Bibliographical Notes

The Schoenberg operators for SVFs in terms of the de Boor algorithm with the metric average are studied in [48, 67]. These works also investigate the Bernstein operators for SVFs in terms of the de Casteljau algorithm with the metric average.

The results on spline subdivision schemes for compact sets based on the Lane–Riesenfeld algorithm with the metric average are obtained in [42].

In [94] a variant of the metric average is proposed, with a metric property relative to the average of the Hausdorff distance between two sets and that of their convex hulls. Several examples in which this average performs better geometrically than the metric average are presented.

A binary metric average relative to the symmetric-difference metric is designed and analyzed in [57]. This average exhibits good geometric properties, and is used in the adaptation of the interpolatory 4-point subdivision scheme to sets. The so-adapted 4-point scheme is the basis of a method for the approximation of a 3D object from its 2D parallel cross-sections [56].

An average of three sets in $\mathbb{R}$ also relative to the symmetric-difference metric is designed in [64], where a method for the approximation in this metric of a 3D object from its parallel 1D cross-sections at the vertices of a 2D triangulation is developed and analyzed.

Chapter 7

Methods Based on Metric Linear Combinations

In the previous chapter, certain positive approximation operators are adapted to SVFs using the metric average. This approach is unsatisfactory, in particular because we could show the convergence of the so-adapted Bernstein approximation operators only for a rather special case of SVFs with images in $\mathbb{R}$.

In this chapter we adapt to SVFs sample-based linear approximation operators of the form

$$A_\chi f(x) = \sum_{i=0}^{N} c_i(x) f(x_i), \qquad (7.1)$$

replacing linear combinations of numbers by metric linear combinations of sets. The resulting operators have the form

$$A_\chi^M F(x) = \bigoplus_{i=0}^{N} c_i(x) F(x_i). \qquad (7.2)$$

The operator (7.2) is termed the **metric analogue** of (7.1), or a metric operator.

Note that in contrast to the operators in Chapters 4, 5 and 6, the operator (7.1) is not necessarily a positive operator, since the metric linear combinations in (7.2) are well defined for any real coefficients.

As concrete examples of metric positive operators, we again investigate the adaptation (7.2) of the Schoenberg spline operators and the Bernstein polynomial operators. We show that the metric Bernstein operators approximate continuous multifunctions of bounded variation with images in $\mathbb{R}^n$. For Hölder continuous SFVs, the metric Schoenberg operators on a uniform partition provide the same order of approximation as those based on the metric average. Yet the metric operators also approximate the wider class of continuous SVFs.

As examples of non-positive operators, we consider the adaptation of polynomial interpolation operators to SVFs. It is shown that the sequence of the metric interpolants at the zeros of the Chebyshev polynomials approximates Lipschitz continuous SFVs with a rate similar to that in the case of Lipschitz continuous real-valued functions.

We start by investigating a basic metric operator.

7.1. Metric Piecewise Linear Interpolation

Here we define the metric piecewise linear interpolation operator and study its properties. We use these results in the next section to analyze the approximation properties of metric operators of the form (7.2).

For a partition $\chi = \{x_0, x_1, \dots, x_N\} \subset [a, b]$ we denote

$$\delta_i = x_{i+1} - x_i, \quad i = 0, \dots, N - 1.$$

Given a multifunction $F : [a, b] \to K(\mathbb{R}^n)$ let $CH(F|_\chi) = CH(F_0, \dots, F_N)$, with $F_i = F(x_i)$, $i = 0, \dots, N$, and with $CH(F_0, \dots, F_N)$ defined as in Section 2.1.

Definition 7.1.1 The metric piecewise linear interpolant to a multifunction F at a partition χ is the SVF

$$S_\chi^M F(x) = \bigoplus_{i=0}^{N} c_i(x) F(x_i),$$

where $c_i(x)$ is the hat function centered at x_i,

$$c_i(x) = \begin{cases} (x - x_{i-1})/(x_i - x_{i-1}), & x_{i-1} \le x \le x_i, \\ (x_{i+1} - x)/(x_{i+1} - x_i), & x_i \le x \le x_{i+1}, \\ 0, & \text{otherwise.} \end{cases} \quad (7.3)$$

By definition, the multifunction $S_\chi^M F$ has a complete representation

$$S_\chi^M F = \{ s(\chi, \varphi) : \varphi \in CH(F|_\chi) \}, \quad (7.4)$$

where $s(\chi, \varphi)$ is a piecewise linear selection interpolating the data (x_i, f_i), $i = 0, \ldots, N$, and $\varphi = (f_0, \ldots, f_N) \in CH(F|_\chi)$.

First we show that $S_\chi^M F(x)$ is identical to the piecewise linear interpolant based on the metric average defined by

$$S_\chi^{MA} F(x) = F_i \oplus_{c_i(x)} F_{i+1}, \quad x \in [x_i, x_{i+1}].$$

Lemma 7.1.2 *For $F : [a, b] \to K(\mathbb{R}^n)$*

$$S_\chi^{MA} F = S_\chi^M F \tag{7.5}$$

and

$$\mathrm{haus}(F(x), S_\chi^M F(x)) \leq 2 \omega_{[a,b]}(F, |\chi|), \; x \in [a, b]. \tag{7.6}$$

Proof To prove (7.5), we first show that $S_\chi^{MA} F(x) \subset S_\chi^M F(x)$ for any $x \in [a, b]$, and then show the opposite inclusion.

For a fixed $x \in [x_i, x_{i+1}]$ and for any $y \in S_\chi^{MA} F(x)$,

$$y = c_i(x) f_i + (1 - c_i(x)) f_{i+1},$$

for some $(f_i, f_{i+1}) \in \Pi(F_i, F_{i+1})$. Thus there exists a metric chain $\varphi \in CH(F|_\chi)$ such that $\varphi = (f_0, \ldots, f_i, f_{i+1}, \ldots, f_N)$, and $y = s(\chi, \varphi)(x)$. Hence $y \in S_\chi^M F(x)$. Also, it is obvious that for any $x \in [a, b]$ and any $\varphi \in CH(F|_\chi)$, we have that $s(\chi, \varphi)(x) \in S_\chi^{MA} F(x)$, which completes the proof of (7.5).

Now, to prove (7.6), we use (7.5), the metric property of the metric average (2.3) and the triangle inequality for the Hausdorff metric, and obtain for $x \in [x_i, x_{i+1}]$

$$\mathrm{haus}(F(x), S_\chi^M F(x))$$

$$= \mathrm{haus}(F(x), S_\chi^{MA} F(x))$$

$$\leq \mathrm{haus}(F(x), F(x_i)) + \mathrm{haus}(F(x_i), S_\chi^{MA} F(x))$$

$$\leq \omega_{[a,b]}(F, \delta_i) + \mathrm{haus}(F(x_i), F(x_{i+1})) \leq 2\omega_{[a,b]}(F, \delta_i),$$

which concludes the proof of the lemma. $\square$

Next we show that $S_\chi^M F$, and its piecewise linear selections given in (7.4) "inherit" regularity properties of the continuous multifunction F. We start with the easy case of Lipschitz continuous SVFs.

Lemma 7.1.3 *Let $F \in Lip([a,b], L)$, and let χ be a partition of $[a,b]$. Then*

$$S_\chi^M F \in Lip([a,b], L).$$

Proof For $x, z \in [x_j, x_{j+1}]$ the claim of the lemma follows from the metric property of the metric average (2.3).

Without loss of generality assume that $x \in [x_j, x_{j+1}]$ and $z \in [x_k, x_{k+1}]$, where $0 \le j < k \le N - 1$. Using the triangle inequality, (2.3) and the Lipschitz continuity of F, we get

$$\mathrm{haus}(S_\chi^M F(x), S_\chi^M F(z)) \le \frac{x_{j+1} - x}{x_{j+1} - x_j} \mathrm{haus}(F_j, F_{j+1})$$

$$+\mathrm{haus}(F_{j+1}, F_k) + \frac{z - x_k}{x_{k+1} - x_k} \mathrm{haus}(F_k, F_{k+1})$$

$$\le L(x_{j+1} - x + x_k - x_{j+1} + z - x_k) \le L|z - x|,$$

and the claim of the lemma follows. □

Using the observation that for $l = 0, \ldots, N - 1$

$$|s(\chi, \varphi)(x_l) - s(\chi, \varphi)(x_{l+1})| \le \mathrm{haus}(S_\chi^M F(x_l), S_\chi^M F(x_{l+1})), \qquad (7.7)$$

and applying the arguments in the proof of the lemma, we get

Corollary 7.1.4 *Under the conditions of Lemma 7.1.3*

$$s(\chi, \varphi) \in Lip([a,b], L),$$

for any $s(\chi, \varphi)$ in (7.4).

Now we consider the case when F is a general continuous function.

Lemma 7.1.5 *Let $F : [a,b] \to K(\mathbb{R}^n)$ be a continuous SVF. Then for any partition χ of $[a,b]$ and for any $\delta > 0$,*

$$\omega_{[a,b]}(S_\chi^M F, \delta) \le 4\,\omega_{[a,b]}(F, \delta). \qquad (7.8)$$

Proof By definition, for any $\delta > 0$

$$\omega_{[a,b]}(S_\chi^M F, \delta) = \sup\{\mathrm{haus}(S_\chi^M F(x), S_\chi^M F(z)) : |x - z| \le \delta, x, z \in [a,b]\}.$$

In case $x, z \in [x_j, x_{j+1}]$, $|x - z| \le \delta$, we get by (7.5), by property (2.3) of the metric average and by the definition of the modulus of continuity that

$$\text{haus}(S_\chi^M F(x), S_\chi^M F(z)) = \frac{|x - z|}{\delta_j} \text{haus}(F_j, F_{j+1})$$

$$\le \frac{|x - z|}{\delta_j} \omega_{[a,b]}(F, \delta_j). \tag{7.9}$$

By property (1.2) of the modulus of continuity with $\lambda = \frac{\delta_j}{\delta}$ we get

$$\frac{|x - z|}{\delta_j} \omega_{[a,b]}(F, \delta_j) \le \frac{|x - z|}{\delta_j}\left(1 + \frac{\delta_j}{\delta}\right) \omega_{[a,b]}(F, \delta)$$

$$\le \left(\frac{|x - z|}{\delta_j} + \frac{|x - z|}{\delta}\right) \omega_{[a,b]}(F, \delta)$$

$$\le 2\,\omega_{[a,b]}(F, \delta), \tag{7.10}$$

which together with (7.9) implies (7.8) in this case, since $\frac{|x-z|}{\delta_j} \le 1$ and $\frac{|x-z|}{\delta} \le 1$.

In case $x \in [x_j, x_{j+1}]$, $z \in [x_k, x_{k+1}]$, $0 \le j < k \le N - 1$ and $|x - z| \le \delta$, we get by the triangle inequality

$$\text{haus}(S_\chi^M F(x), S_\chi^M F(z))$$

$$\le \text{haus}(S_\chi^M F(x), S_\chi^M F(x_{j+1})) + \text{haus}(S_\chi^M F(x_{j+1}), S_\chi^M F(x_k))$$

$$+ \text{haus}(S_\chi^M F(x_k), S_\chi^M F(z)). \tag{7.11}$$

Now, by the interpolation property of $S_\chi^M F$ and since $|x_k - x_{j+1}| \le \delta$, we have

$$\text{haus}(S_\chi^M F(x_{j+1}), S_\chi^M F(x_k)) \le \omega_{[a,b]}(F, \delta). \tag{7.12}$$

Applying (7.9) and (7.10) to the first and third terms in the right-hand side of (7.11) and using (7.12) for the second term we obtain

$$\text{haus}(S_\chi^M F(x), S_\chi^M F(z))$$

$$= \left\{ \frac{x_{j+1} - x}{\delta_j} + \frac{x_{j+1} - x}{\delta} + 1 + \frac{z - x_k}{\delta_k} + \frac{z - x_k}{\delta} \right\} \omega_{[a,b]}(F, \delta)$$

$$\le \left\{ 3 + \frac{x_{j+1} - x + z - x_k}{\delta} \right\} \omega_{[a,b]}(F, \delta) \le 4\omega_{[a,b]}(F, \delta), \tag{7.13}$$

asserting the claim of the lemma. $\qquad\qquad\square$

We can prove similar properties of the selections (7.4) of $S^M_\chi F$ in case of continuous F only for $|x - z| \leq \min_i \delta_i$, since for $x \in [x_j, x_{j+1}]$ and $z \in [x_k, x_{k+1}]$ we have no bound on $|s(\chi, \varphi)(x_k) - s(\chi, \varphi)(x_{j+1})|$ when $k > j + 1$.

Lemma 7.1.6 *For any $s(\chi, \varphi)$ in (7.4) and any $\delta \leq \min_i \delta_i, 0 \leq i \leq N - 1$*

$$\omega_{[a,b]}(s(\chi, \varphi), \delta) \leq 2\omega_{[a,b]}(F, \delta). \tag{7.14}$$

Proof First, let $x, z \in [x_j, x_{j+1}]$, such that $|x - z| < \delta$. Using (7.7) and arguments similar to those in the proof of (7.9) and (7.10) we get

$$
\begin{aligned}
&|s(\chi, \varphi)(x) - s(\chi, \varphi)(z)| \\
&\leq \frac{|x - z|}{\delta_j} \operatorname{haus}(F_j, F_{j+1}) \\
&\leq \frac{|x - z|}{\delta_j} \omega_{[a,b]}(F, \delta_j) \leq \frac{|x - z|}{\delta_j}\left(1 + \frac{\delta_j}{\delta}\right)\omega_{[a,b]}(F, \delta) \\
&= \left(\frac{|x - z|}{\delta_j} + \frac{|x - z|}{\delta}\right)\omega_{[a,b]}(F, \delta) \leq \frac{2|x - z|}{\delta}\omega_{[a,b]}(F, \delta), \quad (7.15)
\end{aligned}
$$

from which (7.14) easily follows, since $|x - z| < \delta$.

For $x \in [x_{j-1}, x_j]$ and $z \in [x_j, x_{j+1}]$ we get, using (7.15),

$$
\begin{aligned}
&|s(\chi, \varphi)(x) - s(\chi, \varphi)(z)| \\
&\leq |s(\chi, \varphi)(x) - s(\chi, \varphi)(x_j)| + |s(\chi, \varphi)(x_j) - s(\chi, \varphi)(z)| \\
&\leq \frac{2}{\delta}(x_j - x)\omega_{[a,b]}(F, \delta) + \frac{2}{\delta}(z - x_j)\omega_{[a,b]}(F, \delta) \\
&\leq \frac{2}{\delta}(z - x)\omega_{[a,b]}(F, \delta), \tag{7.16}
\end{aligned}
$$

which completes the proof, since $z - x \leq \delta$. □

The condition $|x - z| \leq \min_i \delta_i$ in the last lemma can be alleviated for CBV multifunctions.

Lemma 7.1.7 *Let $F \in CBV[a, b]$. Then for any $s(\chi, \varphi)$ in (7.4),*

$$\omega_{[a,b]}(s(\chi, \varphi), \delta) \leq 2\,\omega_{[a,b]}(F, \delta) + \omega_{[a,b]}(v_F, \delta) \leq 3\,\omega_{[a,b]}(v_F, \delta).$$

Proof Denote $s = s(\chi, \varphi)$. For a given $\delta > 0$, let $x \in [x_j, x_{j+1}]$, $z \in [x_k, x_{k+1}]$, $0 \le j \le k \le N - 1$, such that $|x - z| \le \delta$. Now

$$|s(x) - s(z)| \le |s(x) - s(x_{j+1})| + \sum_{l=j+1}^{k-1} |s(x_{l+1}) - s(x_l)| + |s(z) - s(x_k)|.$$

By (7.15) and (7.7)

$$|s(x) - s(z)| \le \frac{2}{\delta}(x_{j+1} - x)\omega_{[a,b]}(F, \delta)$$

$$+ \sum_{l=j+1}^{k-1} \mathrm{haus}(F(x_{l+1}), F(x_l)) + \frac{2}{\delta}(z - x_k)\omega_{[a,b]}(F, \delta).$$

Since $(x_{j+1} - x) + (z - x_k) < \delta$, and by the definition of variation we obtain

$$|s(x) - s(z)| \le 2\,\omega_{[a,b]}(F, \delta) + V_{x_{j+1}}^{x_k}(F).$$

Now, by (1.3) $V_{x_{j+1}}^{x_k}(F) = v_F(x_k) - v_F(x_{j+1})$, thus

$$|s(x) - s(z)| \le 2\omega_{[a,b]}(F, \delta) + v_F(z) - v_F(x).$$

The claim of the lemma follows by taking the supremum over $|x - z| \le \delta$ and using (1.5). $\square$

It follows directly from the fact that φ is a metric chain that $V(s(\chi, \varphi), \chi) \le V(F, \chi)$. Since $s(\chi, \varphi)$ is a piecewise linear function with breakpoints at the points of χ,

$$V(s(\chi, \varphi), \chi) = V_{x_0}^{x_N} s(\chi, \varphi), \tag{7.17}$$

where $\chi = \{x_0, \dots, x_N\}$.

By (7.17) and since

$$\mathrm{haus}(S_\chi^M F(x_i), S_\chi^M F(x_{i+1})) \le \sup_{\varphi \in CH(F|_\chi)} V_{x_i}^{x_{i+1}}(s(\chi, \varphi)),$$

we get

Corollary 7.1.8 *For a CBV multifunction F*

$$V_{x_0}^{x_N}(s(\chi, \varphi)) \le V_{x_0}^{x_N}(F), \quad V_{x_0}^{x_N}(S_\chi^M F) \le V_{x_0}^{x_N}(F).$$

7.2. Error Analysis

We use the metric piecewise linear approximation to obtain error estimates for metric linear operators.

Let $A_\chi^M F$ be defined by (7.2) and $S_\chi^M F$ be the metric piecewise linear interpolation multifunction as in Definition 7.1.1. By the interpolation property of $S_\chi^M F$

$$A_\chi^M F \equiv A_\chi^M (S_\chi^M F). \tag{7.18}$$

Moreover, (7.1), (7.4), the definition of the metric linear combination (2.5) and (7.17) imply

$$A_\chi^M F = \{A_\chi s(\chi, \varphi) : \varphi \in CH(F|_\chi)\}. \tag{7.19}$$

Remark 7.2.1

(i) It follows from (7.19) that the metric analogues of two linear operators of the form (7.1), which are identical on single-valued functions, are identical on SVFs. This is in contrast to the adaptation based on the metric average. For example, the adaptation of the Schoenberg operators using the de Boor algorithm yields different operators than the one using the Lane–Riesenfeld algorithm (compare with Remark 1.3.6(i)).

(ii) By (7.18) and (7.19) the approximation properties of A_χ^M depend on the way A_χ approximates piecewise linear real-valued functions.

First we derive a basic error estimate for metric operators.

Theorem 7.2.2 *Let A_χ be of the form (7.1), and let $F : [a, b] \to K(\mathbb{R}^n)$ be continuous. Then for $x \in [a, b]$*

$$\mathrm{haus}(A_\chi^M F(x), F(x))$$

$$\leq \sup\{|A_\chi s(\chi, \varphi)(x) - s(\chi, \varphi)(x)| : \varphi \in CH(F|_\chi)\}$$

$$+ 2\,\omega_{[a,b]}(F, |\chi|). \tag{7.20}$$

Proof By the triangle inequality and by (7.18)

$$\mathrm{haus}(A_\chi^M F(x), F(x))$$

$$\leq \mathrm{haus}(A_\chi^M (S_\chi^M F)(x), S_\chi^M F(x)) + \mathrm{haus}(S_\chi^M F(x), F(x)),$$

while by (7.19)

$$\mathrm{haus}(A_\chi^M (S_\chi^M F)(x), S_\chi^M F(x)) \leq \sup_{\varphi \in CH(F|_\chi)} |A_\chi s(\chi, \varphi)(x) - s(\chi, \varphi)(x)|.$$

This together with (7.6) completes the proof.   □

The last general theorem guarantees that the metric analogues of operators of the form (7.1), approximating continuous real-valued functions, also approximate continuous SFVs. To get error estimates in terms of the regularity properties of the approximated SVFs, we apply the theorem to three families of SVFs: Lipschitz continuous SVFs, CBV multifunctions and general continuous SVFs.

In what follows $\mathcal{P}L_\chi$ denotes the set of piecewise linear continuous single-valued functions, with values in $\mathbb{R}^n$ and knots at χ.

For Lipschitz continuous SVFs we get in view of Corollary 7.1.4

Corollary 7.2.3 *Let* $F \in Lip([a,b], L)$ *and let* A_χ *be of the form* (7.1), *satisfying*

$$|A_\chi s(x) - s(x)| \leq C\, L\phi(x, |\chi|), \quad s \in \mathcal{P}L_\chi \cap Lip([a,b], L),$$

with ϕ *as in* (1.15).
   Then

$$\mathrm{haus}(A_\chi^M F(x), F(x)) \leq 2\, L|\chi| + CL\phi(x, |\chi|). \tag{7.21}$$

For CBV multifunctions we get a weaker approximation result, by applying Lemma 7.1.7 instead of Corollary 7.1.4.

Corollary 7.2.4 *Let* $F \in CBV[a,b]$, *and let* A_χ *be of the form* (7.1), *satisfying*

$$|A_\chi s(x) - s(x)| \leq C\, \omega_{[a,b]}(s, \phi(x, |\chi|)), \quad s \in \mathcal{P}L_\chi. \tag{7.22}$$

Then

$$\mathrm{haus}(A_\chi^M F(x), F(x)) \leq 2\, \omega_{[a,b]}(F, |\chi|) + 3C\omega_{[a,b]}(v_F, \phi(x, |\chi|)).$$

For continuous SVFs which are not of bounded variation we can prove an approximation result only for a limited class of linear operators, defined on uniform partitions. This class does not contain the Bernstein operators, but contains the Schoenberg operators.

Corollary 7.2.5 *Let* $F : [a,b] \to K(\mathbb{R}^n)$ *be continuous, and let* A_N *be a linear operator of the form* (7.1), *defined on a uniform partition* χ_N *with*

$|\chi_N| = (b-a)/N$, *and satisfying*

$$|A_N s(x) - s(x)| \le C\, \omega_{[a,b]}(s, |\chi_N|), \quad s \in \mathcal{PL}_\chi. \tag{7.23}$$

Then

$$\text{haus}(A_N^M F(x), F(x)) \le 2(1+C)\omega_{[a,b]}(F, |\chi_N|).$$

Proof First we substitute in the first term of (7.20) the error estimate (7.23). Then since $\min_i \delta_i = |\chi_N|$, we can estimate $\omega_{[a,b]}(s, \chi_N)$ using Lemma 7.1.6. □

7.3. Multifunctions with Convex Images

For a multifunction $F : [a, b] \to Co(\mathbb{R}^n)$ the metric approximation $A_\chi^M F$ is a multifunction with images which are not necessarily convex, as is shown in Remark 2.1.3. By convexifying $A_\chi^M F$, namely taking $\text{co}(A_\chi^M F(x))$, we get an approximant with convex images. In fact, the convexification of any non-convex approximant of SVF with convex images does not affect the error bound. This follows from the next Lemma.

Lemma 7.3.1 *Let $A \in Co(\mathbb{R}^n)$ and $B \in K(\mathbb{R}^n)$, then*

$$\text{haus}(A, \text{co}(B)) \le \text{haus}(A, B). \tag{7.24}$$

Proof Obviously

$$\max_{a \in A} \text{dist}(a, \text{co}(B)) \le \max_{a \in A} \text{dist}(a, B) \le \text{haus}(A, B). \tag{7.25}$$

Now, any $b \in \text{co}(B)$ is of the form $b = tb_1 + (1-t)b_2$, $t \in [0, 1]$ with $b_1, b_2 \in B$. Therefore

$$\text{dist}(b, A) \le t\, \text{dist}(b_1, A) + (1-t)\text{dist}(b_2, A),$$

and we get

$$\max_{b \in \text{co}(B)} \text{dist}(b, A) \le \text{haus}(A, B). \tag{7.26}$$

The claim (7.24) follows from (7.25) and (7.26). □

A direct conclusion from the last lemma is

Corollary 7.3.2 *For $F : [a, b] \to Co(\mathbb{R}^n)$ and A any adapted operator to SVFs*

$$\text{haus}(\text{co}(AF(x)), F(x)) \le \text{haus}(AF(x), F(x)). \tag{7.27}$$

Thus the error estimates in Corollaries 7.2.3, 7.2.4 and 7.2.5 also apply for co$(A_\chi^M F)$. Property 4 of the metric linear combination (see Section 2.1.4) guarantees that

$$\text{co}(A_\chi^M F(x)) \subseteq A_\chi^{Mink} F(x), \quad x \in [a, b], \tag{7.28}$$

and hence the convexified metric approximant is "smaller" than $A_\chi^{Mink} F$.

To conclude, we emphasize that co$(A_\chi^M F)$ is a convex-valued approximant of a convex-valued F, which is well defined even for non-positive operators. The error bounds for these approximants follow from the general error analysis for metric approximations and (7.27).

It should be mentioned that an adaptation of interpolation operators to convex-valued multifunctions was proposed by Lempio. In his approach the operators are applied to the representing support functions, and the resulting interpolants are then convexified. To ensure that the so-obtained functions represent a multifunction with non-empty convex images, additional restrictive conditions on the approximated SVF have to be imposed.

7.4. Specific Metric Operators

In this section we present metric analogues of two families of positive operators, and of polynomial interpolation operators, which are not positive operators. We give approximation results, illustrated by examples.

7.4.1. Metric Bernstein operators

Definition 7.4.1 For $F : [0,1] \to K(\mathbb{R}^n)$ the metric Bernstein operator of degree N is

$$B_N^M F(x) = \bigoplus_{i=0}^{N} \binom{N}{i} x^i (1-x)^{N-i} F\left(\frac{i}{N}\right)$$

$$= \left\{ \sum_{i=0}^{N} \binom{N}{i} x^i (1-x)^{N-i} f_i \; : \; (f_0, \ldots, f_N) \in CH(F|_{\chi_N}) \right\},$$

where $CH(F|_{\chi_N}) = CH(F(0), F(1/N), \ldots, F(1))$.

In contrast to the adaptation based on the metric average, we can show for the adaptation based on metric linear combinations that the sequence of so-adapted Bernstein approximants converges as $N \to \infty$ to the approximated CBV multifunctions.

For the case of Lipschitz continuous set-valued functions Corollary 7.2.3 and (1.21) lead to

Corollary 7.4.2 *Let $F \in \text{Lip}([0,1], L)$, then for any $x \in [0,1]$*

$$\text{haus}(B_N^M F(x), F(x)) \leq 2L/N + CL\sqrt{x(1-x)/N}.$$

Using Corollary 7.2.4 and (1.21), we get a more general result for CBV multifunctions.

Corollary 7.4.3 *Let $F \in CBV[0,1]$, then for any $x \in [0,1]$*

$$\text{haus}(B_N^M F(x), F(x)) \leq 2\,\omega_{[0,1]}(F, 1/N) + 3C\omega_{[0,1]}(v_F, \sqrt{x(1-x)/N}).$$

As noted before, we cannot get a similar result for continuous SVFs, since (7.23) does not hold for the Bernstein operators. Yet (7.23) holds for the Schoenberg operators, and therefore they approximate continuous SFVs.

7.4.2. *Metric Schoenberg operators*

Definition 7.4.4 The metric Schoenberg operator of order m for a set-valued function $F : [0,1] \to K(\mathbb{R}^n)$ is defined by

$$S_{m,N}^M F(x) = \bigoplus_{i=0}^{N} b_m(Nx - i)F\left(\frac{1}{N}\right)$$

$$= \left\{ \sum_{i=0}^{N} b_m(Nx - i)f_i : (f_0, \ldots, f_N) \in CH(F|_{\chi_N}) \right\},$$

where $CH(F|_{\chi_N}) = CH(F(0), F(1/N), \ldots, F(1))$.

By (1.26) the Schoenberg operators satisfy (7.23). Thus Corollary 7.2.5 applies, and we get

Corollary 7.4.5 *Let F be a continuous SFV defined on $[0,1]$. Then for any $x \in [\frac{m-1}{N}, 1]$*

$$\text{haus}(S_{m,N}^M F(x), F(x)) = 2\left(1 + \left\lfloor \frac{m+1}{2} \right\rfloor\right)\omega_{[0,1]}(F, 1/N),$$

with $\lfloor t \rfloor$ the maximal integer not greater than t.

The result in Corollary 7.2.4 for CBV multifunctions yields an error estimate inferior to the one in Corollary 7.4.5. Yet in the specific case of Lipschitz continuous SVFs, the error bound can be further improved by applying Corollary 7.2.3.

Corollary 7.4.6 *Let $F \in \text{Lip}([0,1], L)$, then for any $x \in [\frac{m-1}{N}, 1]$*

$$\text{haus}(S_{m,N}^M F(x), F(x)) = \left(2 + \left\lfloor \frac{m+1}{2} \right\rfloor\right) \frac{L}{N}.$$

7.4.3. *Metric polynomial interpolation*

Here we study the metric analogues of polynomial interpolation operators, which are not positive operators. Approximation results and examples are presented. The two graphical examples we present assert that such interpolation between sets is reasonable from the geometric point of view.

Definition 7.4.7 For given data (x_i, A_i), $i = 0, \ldots, N$, where $x_0, \ldots, x_N$ are distinct real points and $A_i \in K(\mathbb{R}^n)$, we define the metric polynomial interpolant to this data by

$$\bigoplus_{i=0}^{N} l_i(x) A_i,$$

with

$$l_i(x) = \prod_{j=0, j \neq i}^{N} \frac{x - x_j}{x_i - x_j}, \quad i = 0, \ldots, N.$$

In particular, for $F : [a, b] \to K(\mathbb{R}^n)$, the metric polynomial interpolation operator at the partition χ of $[a, b]$, is given by

$$P_\chi^M F(x) = \bigoplus_{i=0}^{N} l_i(x) F(x_i). \tag{7.29}$$

It is known that for a Lipschitz continuous real-valued function f, its sequence of polynomial interpolants at the Chebyschev roots converges to f. Next we extend this result to SVFs using Corollary 7.2.3 and (1.31).

Corollary 7.4.8 *Let $F \in \text{Lip}([-1,1], L)$, and let the points of the partition χ be the roots of the Chebyshev polynomial of degree $N + 1$ on $[-1,1]$, then*

$$\text{haus}(P_\chi^M F(x), F(x)) \leq 2L|\chi| + \frac{C \log N}{N} = O\left(\frac{\log N}{N}\right), \quad x \in [-1,1].$$

The last equality is obtained by the observation $|\chi| \leq \frac{\pi}{N}$, which follows from the explicit form of the Chebyshev roots $x_i = \cos\left(\frac{2i-1}{2N}\pi\right)$, $i = 0, \ldots, N$. In fact

$$\left| \cos\left(\frac{2i-1}{2N}\pi\right) - \cos\left(\frac{2(i-1)-1}{2N}\pi\right) \right|$$

$$= 2\left| \sin\left(\frac{\pi(i-1)}{N}\right) \right| \left| \sin\left(\frac{\pi}{2N}\right) \right| \leq 2\left| \sin\left(\frac{\pi}{2N}\right) \right| \leq 2\left| \frac{\pi}{2N} \right| = \frac{\pi}{N}.$$

Thus the sequence of metric polynomial interpolants at the roots of the Chebyshev polynomials tends to the Lipschitz continuous interpolated SVF as $N \to \infty$.

Two examples of metric parabolic interpolation to three sets.

To illustrate the metric set-valued polynomial interpolation, and to see the geometry of metric linear combinations of sets with negative coefficients (as occurs in interpolation), we present two examples of a metric parabolic interpolant to three sets in R.

In the first example the interpolated data (x_i, A_i), $i = 0, 1, 2$, is

$$(1, [2, 8]) \quad (4, \{6.5\}) \quad (9, \{8.5\}),$$

while in the second example it is:

$$(0, [2, 4] \cup [6, 8]) \quad (4, [4.5, 5.5]) \quad (8, [2, 4] \cup [6, 8]).$$

The two set-valued interpolants are illustrated in Figs. 7.4.9 and 7.4.10, respectively. In the figures the interpolated sets are depicted in black. The gray curves in the left-hand side figures are parabolic interpolants to some metric chains of the three sets. The figures in the right-hand side depict in gray the graphs of the set-valued interpolants (the union of the parabolic interpolants to all metric chains).

Figure 7.4.9   Metric parabolic interpolant — first example.

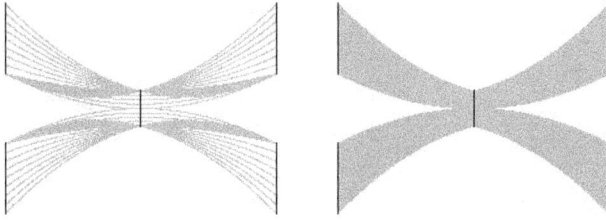

Figure 7.4.10   Metric parabolic interpolant — second example.

7.5.   Bibliographical Notes

The results in this chapter are based on [44, 67]. Another adaptation of positive sample-based operators is constructed in [58], based on a new average of several sets. Approximation by such operators of univariate and multivariate SVFs in the symmetric-difference metric are studied there.

Non-positive interpolation operators for SVFs with convex images are investigated by Lempio [63]. Approximation of multifunctions with convex images by non-positive operators is also possible through the embedding of $Co(\mathbb{R}^n)$ in the Banach space of directed sets [12, 13]. In this approach the values of the approximants are not necessarily convex sets, but differences of two convex sets embedded in this space. Specific approximants within the space of directed sets are studied in [11, 17, 79].

Chapter 8

Methods Based on Metric Selections

In this chapter we consider a complete representation of a CBV multifunction F by specific selections with uniformly bounded variation. This follows from the important property of a CBV multifunction is that through any point in $Graph(F)$ there passes a CBV selection, with variation bounded by that of F.

Here we obtain this known result using the construction of metric chains, and therefore we term these selections "metric selections". In addition, our method of proof provides an estimate of the modulus of continuity of the metric selections. We use the representation by metric selections to adapt general approximation operators to CBV multifunctions by applying the operators to the metric selections, and obtain error bounds in terms of the regularity properties of the approximated SVF.

This type of adaptation applies to any complete representation $\mathcal{R}$ of a multifunction F, and yields error bounds in terms of the regularity properties of the selections in $\mathcal{R}$. In case these selections are C^k for some $k \geq 1$ the error estimates in the Hausdorff metric can be expressed in terms of the k-th modulus of smoothness of F based on $\mathcal{R}$, as defined in Section 3.3.

8.1. Metric Selections

Here we show that through any point of the graph of $F \in CBV[a, b]$ there exists a continuous selection, and we derive its regularity properties in terms of those of F.

Theorem 8.1.1 *Let $F : [a, b] \to K(\mathbb{R}^n)$ be a CBV multifunction. Then through any point in Graph(F) there exists a continuous selection s satisfying*

$$V_a^b(s) \leq V_a^b(F), \tag{8.1}$$

$$\omega_{[a,b]}(s, \delta) \leq 3\,\omega_{[a,b]}(v_F, \delta), \quad \delta > 0. \tag{8.2}$$

In particular if $F \in Lip([a, b], L)$, then (8.2) can be replaced by

$$s \in Lip([a, b], L). \tag{8.3}$$

Proof For a fixed $(\widetilde{x}, \widetilde{y}) \in Graph(F)$, $\widetilde{x} \in [a, b]$, $\widetilde{y} \in F(x)$ we construct metric chains. Let $x_i = a + ih$, $i = 0, \ldots, N$, $Nh = b - a$ and let j be be such that $x_j \leq \widetilde{x} \leq x_{j+1}$. Choose

$$y_j \in \Pi_{F(x_j)}(\widetilde{y}), \quad y_k \in \Pi_{F(x_k)}(y_{k+1}), \quad k = j - 1, j - 2, \ldots, 0,$$

and similarly

$$y_{j+1} \in \Pi_{F(x_{j+1})}(\widetilde{y}), \quad y_k \in \Pi_{F(x_k)}(y_{k-1}), \quad k = j + 2, \ldots, N.$$

Now we define the partition $\widetilde{\chi}_N = \{x_0, \ldots, x_j, \widetilde{x}, x_{j+1}, \ldots, x_N\}$. For the metric chain $\varphi = \{y_0, \ldots, y_j, \widetilde{y}, y_{j+1}, \ldots, y_N\}$ let s_N be the piecewise linear function interpolating the points $(\widetilde{x}, \widetilde{y})$, (x_i, y_i), $i = 0, \ldots, N$. It follows from Lemma 7.1.7 that for all $N \in \mathbb{N}$

$$\omega_{[a,b]}(s_N, \delta) \leq 3\,\omega_{[a,b]}(v_F, \delta), \quad \delta > 0, \tag{8.4}$$

while by Corollary 7.1.4 for $F \in Lip([a, b], L)$

$$s_N \in Lip([a, b], L). \tag{8.5}$$

The sequence of piecewise linear functions $\{s_N : N \in \mathbb{Z}_+\}$ is equicontinuous by (8.4) or by (8.5) and equibounded since $Graph(F)$ is bounded. Then by the Arzela–Ascoli theorem there exists a subsequence which converges to a continuous function s passing through $(\widetilde{x}, \widetilde{y})$. It is easy to see that s is a selection of F, since $Graph(F)$ is closed.

Next we prove the properties of the selection s. By Corollary 7.1.8

$$V_a^b(s_N) \leq V_a^b(F), \quad N \in \mathbb{N}. \tag{8.6}$$

Since any s_N satisfies (8.6), and (8.4) or (8.5), we conclude that the limit function satisfies (8.1), and (8.2) or (8.3). $\qquad \square$

We term any selection through a point α in $Graph(F)$ guaranteed by Theorem 8.1.1 **metric selection**, and denote it by f^α.

Corollary 8.1.2 *Any multifunction $F \in CBV[a,b]$ has a complete representation by metric selections*

$$F = \{f^\alpha : \alpha \in Graph(F)\} \tag{8.7}$$

where f^α satisfies (8.1) and (8.2). In the particular case $F \in Lip([a,b], L)$, (8.2) can be replaced by (8.3).

Note that the representation (8.7) is not necessarily unique.

By Lemma 3.3.1 and (8.2) we have

$$\omega_{[a,b]}(F, \delta) \leq \sup_{\alpha \in Graph(F)} \omega_{[a,b]}(f^\alpha, \delta) \leq 3\omega_{[a,b]}(v_F, \delta).$$

This implies that for a CBV multifunction F, the continuity of its metric selections is characterized by that of F and its variation function v_F.

It is easy to conclude from the definition of a metric selection f^α of F that if $(\tilde{x}, f^\alpha(\tilde{x}))$ is an interior point of $Graph(F)$ then there exists a neighborhood of $\tilde{x}$ such that $f^\alpha(x)$ is constant in this neighborhood.

The following example illustrates various types of metric selections.

Example 8.1.3 Consider the multifunction $F : [-2,2] \to K(\mathbb{R})$ given by

$$F(x) = \begin{cases} [-2, 2], & 1 < |x| \leq 2, \\ [-2, -\sqrt{1-x^2}] \cup [\sqrt{1-x^2}, 2], & |x| \leq 1. \end{cases}$$

Here are three main types of metric selections f^α, passing through the point $\alpha = (\alpha_1, \alpha_2) \in Graph(F)$.

(1) $f^\alpha(x) = \alpha_2$, $|\alpha_2| \geq 1$;

(2) $f^\alpha(x) = \begin{cases} \alpha_2, & x \in [-2, -\sqrt{1-\alpha_2^2}), \\ \sqrt{1-x^2}, & x \in [-\sqrt{1-\alpha_2^2}, 0), \quad \alpha_1 < 0, 0 \leq \alpha_2 < 1; \\ 1, & x \in [0, 2], \end{cases}$

(3) $f^\alpha(x) = \begin{cases} \alpha_2, & x \in [-2, -\sqrt{1-\alpha_2^2}), \\ -\sqrt{1-x^2}, & x \in [-\sqrt{1-\alpha_2^2}, 0), \quad \alpha_1 < 0, -1 < \alpha_2 \leq 0. \\ -1, & x \in [0, 2], \end{cases}$

For $\alpha_1 > 0$ and $-1 \leq \alpha_2 \leq 1$ we get two more types of metric selections which are reflections with respect to $x = 0$ of the metric selections (2) and (3).

Some of the metric selections of F are illustrated below.

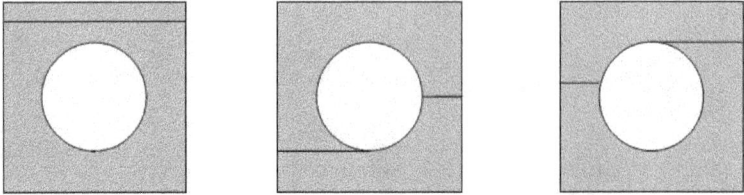

Figure 8.1.4 Some metric selections of F.

In the following section we use the complete representations by metric selections to approximate SFVs.

8.2. Approximation Results

Using the representation introduced in Corollary 8.1.2 we adapt approximation operators defined on real-valued functions to SVFs by applying them to the metric selections.

Definition 8.2.1 Let $F \in CBV[a,b]$ with a representation (8.7) and let A be an operator defined on continuous real-valued functions. We adapt A to F by

$$A^{MS}F(x) = \{Af^\alpha(x) : \alpha \in Graph(F)\}, \quad x \in [a,b].$$

Obviously $A^{MS}F$ depends on the representation (8.7). Yet, as shown in the next theorem, all such approximations have the same error bound.

Theorem 8.2.2 *Let A_δ be an approximation operator satisfying for any $x \in [a,b]$*

$$|A_\delta f(x) - f(x)| \leq C\,\omega_{[a,b]}(f, \phi(x,\delta)), \quad f \in C[a,b],$$

with δ a small parameter and with ϕ as in (1.15).

Then for $F \in CBV[a,b]$ and for a complete representation of F by metric selections

$$\mathrm{haus}(F(x), A_\delta^{MS}F(x)) \leq 3C\omega_{[a,b]}(v_F, \phi(x,\delta)), \quad x \in [a,b]. \tag{8.8}$$

In particular, if $F \in Lip\,[a,b]$ then

$$\text{haus}(F(x), A_\delta^{MS} F(x)) \leq CL\phi(x,\delta), \quad \in [a,b]. \tag{8.9}$$

Proof Given a representation of F, $\{f^\alpha : \alpha \in \mathcal{A}\}$, for any $(x,y) \in Graph(F)$ there exists α_1 such that $y = f^{\alpha_1}(x)$. Then we have

$$\text{dist}(y, A_\delta^{MS} F(x)) = \text{dist}(f^{\alpha_1}(x), A_\delta^{MS} F(x))$$

$$\leq |f^{\alpha_1}(x) - A_\delta f^{\alpha_1}(x)| \leq C\omega_{[a,b]}(f^{\alpha_1}, \phi(x,\delta)).$$

Therefore using (8.2) we obtain

$$\sup_{y \in F(x)} \text{dist}(y, A_\delta^{MS} F(x)) \leq 3C\omega_{[a,b]}(v_F, \phi(x,\delta)). \tag{8.10}$$

Now, let $a \in A_\delta^{MS} F(x)$. By Definition 8.2.1 there exists $A_\delta f^{\alpha_2}$ such that $a = A_\delta f^{\alpha_2}(x)$, where f^{α_2} is a selection in the representation $F = \{f^\alpha : \alpha \in Graph(F)\}$. Then

$$\text{dist}(a, F(x)) = \text{dist}(A_\delta f^{\alpha_2}(x), F(x))$$

$$\leq |A_\delta f^{\alpha_2}(x) - f^{\alpha_2}(x)| \leq C\omega_{[a,b]}(f^{\alpha_2}, \phi(x,\delta)).$$

Thus by (8.2) we get

$$\sup_{a \in A_\delta F(x)} \text{dist}(a, F(x)) \leq C \sup_{\alpha \in Graph(F)} \omega_{[a,b]}(f^\alpha, \phi(x,\delta))$$

$$\leq 3C\omega_{[a,b]}(v_F, \phi(x,\delta)). \tag{8.11}$$

Equations (8.10) and (8.11) imply (8.8).

If $F \in Lip([a,b], L)$, then we can improve the last estimate in (8.10) and in (8.11) using (8.3) in Theorem 8.1.1. Thus

$$\text{haus}(F(x), A_\delta F(x)) \leq CL\phi(x,\delta),$$

which is (8.9). $\qquad\qquad\qquad\qquad\qquad\qquad\qquad\qquad\qquad\qquad\square$

As a byproduct of the proof of the last theorem, and in view of (8.3) and (1.31), we get

Corollary 8.2.3 *Let $F \in Lip([a,b], L)$ and let P_χ be the polynomial interpolation operator (1.29) at the points χ, which are the roots of the Chebyshev polynomial of degree $N+1$. Then*

$$\text{haus}(F(x), P_\chi^{MS} F(x)) \leq \frac{C\log N}{N}.$$

Remark 8.2.4

(i) Comparing the adaptation based on metric selections with that based on metric linear combinations, we observe

   (a) The first adaptation method applies to general approximation operators defined on real-valued functions, while the second adaptation method is limited to linear sample-based operators.
   (b) The error bounds in Theorem 8.2.2 for the first adaptation are sharper than the corresponding ones for the second adaptation in Corollaries 7.2.3 and 7.2.4, where in the error bounds there is an additional term $2\omega_{[a,b]}(F, |\chi|)$.

(ii) Applying Corollary 7.3.2 to the approximants based on metric selections, we get convex approximants to convex-valued multifunctions, which in the case of positive linear sample-based operators have similar error bounds as for the adaptation based on Minkowski convex combinations (presented in Section 4.3).

(iii) It is not hard to realize that a general CBV multifunction F has non-smooth metric selections, even if the boundaries of the graph of F are smooth. Figure 8.1.4 demonstrates this observation. Thus F is not $\mathcal{R}$-smooth of a positive order, with $\mathcal{R}$ a representation of F by metric selections (see Section 3.3).

(iv) From the proof of Theorem 8.2.2 it follows that for any complete represenation $\mathcal{R}$ of F, $F(x) = \{f^\xi(x) : \xi \in \Xi\}$, $x \in [a,b]$, the adapted operator $A^\mathcal{R}$ satisfies

$$\text{haus}(F(x), A^\mathcal{R} F(x)) \le \sup_{\xi \in \Xi} \|f^\xi(x) - Af^\xi(x)\|_\infty. \qquad (8.12)$$

Thus for C^k selections in $\mathcal{R}$, with equicontinuous k-th derivatives, error bounds of high order can be obtained. In view of (iii) in this remark, such error estimates are rare in the case of metric selections.

8.3. Bibliographical Notes

The results in this chapter are based on [67]. Previous results related to regular selections of continuous multifunctions of bounded variation and of Lipschitz SVFs may be found in [33, 53]. Special investigation of multifunctions of bounded variation and their selections is done in [23–25]. In particular the existence of a CBV selection through any point in the graph of a CBV multifunction is proved in [23]. More about regular selections can be found in the bibliographical notes of Chapter 3.

PART III

Approximation of SVFs with Images in $\mathbb{R}$

Chapter 9

SVFs with Images in $\mathbb{R}$

In this part of the book we study special methods of representation and approximation of SVFs with images in $K(\mathbb{R})$. Since compact sets in $\mathbb{R}$ are much simpler than compact sets in $\mathbb{R}^n$ with $n \geq 2$, simple representations of SVFs with images in $K(\mathbb{R})$ can be constructed, yielding efficient approximation methods.

In Section 4.4 we discuss a canonical representation of SVFs with images in $K(\mathbb{R})$, based on the parametrization (2.13). Although being rather simple, this representation does not lead to satisfactory approximation methods.

In the next chapter we study several simple representations of a certain subclass of CBV multifunction with images in $K(\mathbb{R})$, by a collection of special selections which inherit regularity properties from those of the SVF. These representations are used in Chapter 11 to adapt approximation operators to such multifunctions.

To obtain the above representations, in this chapter we analyze the graphs of CBV multifunctions. In particular we prove the continuity of the boundaries of their graphs, considered as real-valued functions, and investigate their regularity properties in relation to those of the SVFs. The restriction of the analysis to CBV multifunctions guarantees the continuity of the special selections, constructed in the next chapter. The main tool in our analysis is the existence of a continuous selection through any points of the graph (Theorem 8.1.1).

The bibliographical notes of the three chapters of this part are given in Section 11.3.

9.1. Preliminaries on the Graphs of SVFs

We present specific notions and definitions related to the graphs of multifunctions with images in $K(\mathbb{R})$. In particular we introduce the notion of a point of change of topology (PCT). We single out a special type of PCTs, called singular PCTs, and show that the absence of such points is necessary for the continuity of SVFs in the Hausdorff metric.

Let $F : [a, b] \to K(\mathbb{R})$, and recall that

$$(\mathrm{co}F)(x) = \mathrm{co}(F(x)), \quad x \in [a, b].$$

Now consider the set

$$Graph(\mathrm{co}F) \setminus Graph(F). \tag{9.1}$$

We call a maximal connected open subset of (9.1) a **hole of F**. The collection of all such holes is denoted by $\mathcal{H}(F)$. The number of holes in $\mathcal{H}(F)$ is denoted by $|\mathcal{H}(F)|$.

The boundary of a hole $H \in \mathcal{H}(F)$ in the graph of F is

$$\widetilde{\partial}H = \mathrm{cl}(H) \bigcap Graph(F).$$

An **interior hole** of F is a hole H for which $\mathrm{cl}(H) \setminus H = \widetilde{\partial}H$. All other holes are termed **boundary holes**. In Fig. 9.1.2 the four holes with boundaries containing the points $p_3, \ldots, p_{10}$ are interior holes; the other two holes are boundary holes.

The domain of a hole H is denoted by $\Delta_H = [x_H^l, x_H^r]$ with

$$x_H^l = \inf\{x : \exists y \text{ such that } (x, y) \in H\},$$
$$x_H^r = \sup\{x : \exists y \text{ such that } (x, y) \in H\}.$$

For $x_H^l < x < x_H^r$, we call the set $H(x) = \{y : (x, y) \in H\}$ the cross-section of H at x. Note that $H(x)$ is a non-empty open set since H is connected and open.

Points in $Graph(F)$, where locally the topology of the images of F changes, play a central role in our analysis.

Definition 9.1.1 A point $(x, y) \in Graph(F)$ is called a **point of change of topology (PCT)** of F if for any $\varepsilon > 0$ small enough there exists $\delta(\varepsilon) \in \mathbb{R}$, $\delta(\varepsilon) \neq 0$ such that $\forall z$ satisfying

$$\min\{0, \delta(\varepsilon)\} < z - x < \max\{0, \delta(\varepsilon)\}, \tag{9.2}$$

the two sets $F(x) \cap \mathcal{B}(y, \varepsilon)$ and $F(z) \cap \mathcal{B}(y, \varepsilon)$ consist of a different number of intervals (a single point is considered as an interval of zero length). Here $\mathcal{B}(y, \varepsilon)$ is the interval $(y - \varepsilon, y + \varepsilon)$.

It is easy to see that any point of change of topology p is associated with some hole $H \in \mathcal{H}(F)$, namely $p \in \widetilde{\partial}H$.

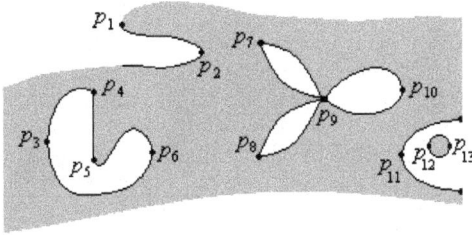

Figure 9.1.2   Graph of F with holes and PCTs.

In Fig. 9.1.2 the graph of a multifunction F with images in $\mathbb{R}$ is shown in gray. The collection $\{\widetilde{\partial}H : H \in \mathcal{H}(F)\}$ is depicted in black. The points p_1, p_2, p_3, all the points on $(p_4, p_5]$ and the points $p_6, \ldots, p_{13}$ are PCTs of F. Note that the points p_1, p_2 and p_{11}, p_{12}, p_{13} are associated with boundary holes, while the rest are associated with interior holes.

Among the points of change of topology we mark out the "singular" ones which do not exist in graphs of continuous SVFs.

Definition 9.1.3 A point of change of topology (x, y) is called **singular** if

$$F(z) \bigcap \mathcal{B}(y, \varepsilon) = \emptyset,$$

with ε and z as in Definition 9.1.1.

It follows directly from Definition 9.1.3 that any isolated point of $Graph(F)$ is a singular PCT. (Recall that p is an isolated point of a set S, if one can find an open ball $\mathcal{B}(p, \varepsilon)$ which contains no other points of S.)

In Fig. 9.1.2 the points p_1, p_{12}, p_{13} and all points of $[p_4, p_5] \setminus \{p_4\}$ are singular PCTs.

Lemma 9.1.4 *A continuous multifunction F does not have singular PCTs.*

Proof Suppose (x, y), $y \in F(x)$ is a singular PCT. By Definition 9.1.3 there exist $\varepsilon > 0$ and $\delta(\varepsilon) \neq 0$ such that for every z satisfying (9.2), dist $(y, F(z)) \geq \varepsilon$. Then haus$(F(x), F(z)) \geq \varepsilon$, implying that F is discontinuous at x. $\qquad\qquad\square$

In the next section we restrict the discussion to continuous multifunctions of bounded variation, and show that the boundaries of their graphs are continuous. The requirement that the multifunction is CBV is essential, as is demonstrated by the simple example

$$F(x) = \begin{cases} \left[-2, \sin\dfrac{1}{x}\right] \bigcup \left[\dfrac{|x|}{2} + \sin\dfrac{1}{x}, 2\right], & x \in [-2, 2] \setminus \{0\}, \\ [-2, 2], & x = 0. \end{cases} \tag{9.3}$$

Note that F is continuous in the Hausdorff metric, yet $x = 0$ is a discontinuity point of the boundary of the holes of F.

9.2.   Continuity of the Boundaries of a CBV Multifunction

Here we show that the boundaries of the graph of a CBV multifunction are continuous, with regularity properties related to those of the multifunction.

First we define the **lower boundary** and the **upper boundary** of a multifunction $F : [a, b] \to K(\mathbb{R})$ as follows

$$\begin{aligned} f^{low}(x) &= \min\{y : y \in F(x)\}, & x \in [a, b], \\ f^{up}(x) &= \max\{y : y \in F(x)\}, & x \in [a, b], \end{aligned} \tag{9.4}$$

and then show that for a continuous multifunction F, f^{low} and f^{up} are continuous.

Theorem 9.2.1 *Let $F : [a, b] \to K(\mathbb{R})$ be a continuous SVF. Then f^{low} and f^{up} are continuous.*

Proof In the following, f is either f^{low} or f^{up}. Note that by (9.4), for any $x, z \in [a, b]$, $|f(x) - f(z)|$ is either $\text{dist}(f(x), F(z))$ or $\text{dist}(f(z), F(x))$. Hence

$$|f(x) - f(z)| \leq \text{haus}(F(x), F(z)), \tag{9.5}$$

which implies the claim of the theorem.   □

Next, for any hole $H \in \mathcal{H}(F)$ we define its **lower boundary** and **upper boundary** as follows

$$\begin{aligned} b_H^{low}(x) &= \inf\{y : (x, y) \in H\}, & x \in (x_H^l, x_H^r) \\ b_H^{up}(x) &= \sup\{y : (x, y) \in H\}, & x \in (x_H^l, x_H^r). \end{aligned} \tag{9.6}$$

These functions, together with f^{low} and f^{up}, are termed the **boundaries of F**. We use the notation

$$\partial F = \{b_H^{low}, b_H^{up} : H \in \mathcal{H}(F)\} \bigcup \{f^{low}, f^{up}\}, \tag{9.7}$$

and denote by D_f the domain of $f \in \partial F$. Note that $Graph(f) \subset Graph(F)$ for any $f \in \partial F$, since $Graph(F)$ is a closed set (see Section 3.1).

To show that all the boundaries of $F \in CBV[a,b]$ are continuous, we prove a series of lemmas. The proofs of these lemmas are based on the existence of a continuous selection through any point in the graph of a CBV multi-function. The proofs are rather intricate using "hard analysis" methods. The reader interested in the approximation results can omit these proofs.

The first lemma shows that all the cross-sections of the holes of a CBV multifunction are convex.

Lemma 9.2.2 *Let* $F \in CBV[a,b]$, *and let* $H \in \mathcal{H}(F)$. *Then for any* $x \in (x_H^l, x_H^r)$ *the cross-section of the hole* H *at* x, $H(x) = \{y : (x,y) \in H\}$, *is an open interval.*

Proof We prove the lemma by contradiction. Since $H(x)$ is open, we assume the contrary, namely that there exists $\widetilde{x} \in (x_H^l, x_H^r)$ such that $H(\widetilde{x})$ is not convex, or equivalently that there exists $(\widetilde{x}, \widetilde{y}) \in Graph(F)$, with $\widetilde{y} \in co(H(\widetilde{x})) \setminus H(\widetilde{x}) \subset F(\widetilde{x})$.

Now, let $s(x)$ be the continuous selection through the point $(\widetilde{x}, \widetilde{y}) \in Graph(F)$ guaranteed by Theorem 8.1.1. Then by definition

$$\{(x, s(x)) : a \le x \le b\} \subset Graph(F). \tag{9.8}$$

Define

$$H_s^{up} = \{(x,y) : \quad (x,y) \in H, \ y > s(x)\}, \tag{9.9}$$

$$H_s^{low} = \{(x,y) : \quad (x,y) \in H, \ y < s(x)\}. \tag{9.10}$$

Obviously $H_s^{up} \cap H_s^{low}$ is empty. By (9.8) $H_s^{up} \cup H_s^{low} = H$. This contradicts the connectivity of H, in view of the continuity of s. $\square$

The following two lemmas lead to the continuity of the boundaries of a hole H on the open interval (x_H^l, x_H^r). The first lemma is rather intuitive.

Lemma 9.2.3 *Let* s *be a continuous selection of* $F \in CBV[a,b]$ *and let* $H \in \mathcal{H}(F)$. *Then either* $s(x) \ge b_H^{up}(x)$, $\forall x \in (x_H^l, x_H^r)$ *or* $s(x) \le b_H^{low}(x)$, $\forall x \in (x_H^l, x_H^r)$.

Proof For $x \in (x_H^l, x_H^r)$, denote $b_x = b_H^{up}(x)$ and assume that $s(x) \ge b_x$. We prove in this case the first inequality. Assume, to the contrary, that for some $z \in (x_H^l, x_H^r)$ $s(z) < b_H^{up}(z)$, and therefore by Lemma 9.2.2 $s(z) \le b_H^{low}(z) = b_z$. By the connectivity of H for a small $\varepsilon > 0$ there exists a continuous path $\varphi \subset H$ from $(x, b_x - \varepsilon)$ to $(z, b_z + \varepsilon)$. Since

s is continuous and passes through the points $(x, s(x))$, $(z, s(z))$, with $s(x) > b_x - \varepsilon$, $s(z) < b_z + \varepsilon$, it must intersect $\varphi \subset H$. This contradicts the definition of a selection.

The second inequality is proved similarly. $\qquad\square$

Lemma 9.2.4 *For $F \in CBV[a, b]$ and $H \in \mathcal{H}(F)$, b_H^{up} and b_H^{low} are continuous on (x_H^l, x_H^r).*

Proof We prove the lemma for the function b_H^{up} by contradiction. Suppose that b_H^{up} is not continuous at $\widetilde{x} \in (x_H^l, x_H^r)$, then there is a sequence $\{x_n\}_{n=1}^{\infty} \subset (x_H^l, x_H^r)$ with $\lim_{n \to \infty} x_n = \widetilde{x}$, such that

$$\lim_{n \to \infty} b_H^{up}(x_n) = \widetilde{y} \neq b_H^{up}(\widetilde{x}). \tag{9.11}$$

There are two possibilities, either $\widetilde{y} > b_H^{up}(\widetilde{x})$ or $\widetilde{y} < b_H^{up}(\widetilde{x})$.

(i) Assume $\widetilde{y} > b_H^{up}(\widetilde{x})$. By Theorem 8.1.1 there exists a continuous selection $s(x)$ through the point $(\widetilde{x}, b_H^{up}(\widetilde{x}))$. We define the sets H_s^{up} and H_s^{low} by (9.9) and (9.10). Since the points $(x_n, b_H^{up}(x_n))$ can be approximated by points from H, it follows that $H_s^{up} \neq \emptyset$. By Lemma 9.2.2 the set $H(\widetilde{x})$ is not empty, therefore $H_s^{low} \neq \emptyset$. Thus we get a contradiction to Lemma 9.2.3.

(ii) Now assume that $\widetilde{y} < b_H^{up}(\widetilde{x})$. For $n = 1, 2, \ldots$ there exists a continuous selection $s_n(x)$ through $(x_n, b_H^{up}(x_n))$, satisfying the claims of Theorem 8.1.1. By Lemma 9.2.3

$$s_n(x) \geq b_H^{up}(x), \quad \forall x \in (x_H^l, x_H^r). \tag{9.12}$$

By (8.2), the selections $\{s_n\}_{n=1}^{\infty}$ are uniformly continuous, hence by the Arzela–Ascoli theorem there exists a uniformly convergent subsequence $\{s_{n_k}\}_{k=1}^{\infty}$. Let $\lim_{k \to \infty} s_{n_k} = \widetilde{s}$. We now show that $\widetilde{s}(\widetilde{x}) = \widetilde{y}$. Indeed, by (8.2)

$$|s_{n_k}(\widetilde{x}) - \widetilde{y}| \leq |s_{n_k}(\widetilde{x}) - s_{n_k}(x_{n_k})| + |s_{n_k}(x_{n_k}) - \widetilde{y}|$$

$$\leq 3\omega_{[a,b]}(v_F, |\widetilde{x} - x_{n_k}|) + |b_H^{up}(x_{n_k}) - \widetilde{y}|, \tag{9.13}$$

and since the function v_F is continuous, we get by (9.11) that

$$\lim_{k \to \infty} s_{n_k}(\widetilde{x}) = \widetilde{y}.$$

Finally we get the contradiction

$$\widetilde{s}(\widetilde{x}) = \lim_{k \to \infty} s_{n_k}(\widetilde{x}) \geq b_H^{up}(\widetilde{x}) > \widetilde{y} = \widetilde{s}(\widetilde{x}).$$

using (9.12) and the assumption $\widetilde{y} < b_H^{up}(\widetilde{x})$.

The proof for b_H^{low} is similar. $\qquad\square$

In the next lemma we extend the continuity of b_H^{up}, b_H^{low} to the closed interval $[x_H^l, x_H^r]$.

Lemma 9.2.5 *For* $F \in CBV[a, b]$ *and* $H \in \mathcal{H}(F)$, *the following limits exist*

$$\lim_{x \to x_H^l} b_H^{up}(x), \quad \lim_{x \to x_H^r} b_H^{up}(x), \quad \lim_{x \to x_H^l} b_H^{low}(x), \quad \lim_{x \to x_H^r} b_H^{low}(x).$$

Proof The proof of the existence of each of the four limits can be done in a similar way. We prove that $\lim_{x \to x_H^l} b_H^{up}(x)$ exists.

Assume, to the contrary, that there are $\{x_k\} \to x_H^l$ and $\{z_k\} \to x_H^l$ such that $\{b_H^{up}(x_k)\} \to f_1$, $\{b_H^{up}(z_k)\} \to f_2$ with $f_1 < f_2$. Consider the point $\tilde{f} = (1/2)f_1 + (1/2)f_2$. By the continuity of b_H^{up} on (x_H^l, x_H^r) there exists $\{\xi_k\}$, $\xi_k \in \mathrm{co}\{x_k, z_k\}$ such that

$$b_H^{up}(\xi_k) = (1/2)b_H^{up}(x_k) + (1/2)b_H^{up}(z_k).$$

Since $\lim_{k \to \infty} b_H^{up}(\xi_k) = \tilde{f}$, and $Graph(F)$ is a closed set, we get $(x_H^l, \tilde{f}) \in Graph(F)$.

Now, let s be a continuous selection through $(x_H^l, \tilde{f})$. Again we define the sets H_s^{up} and H_s^{low} by (9.9) and (9.10). Since the points $(x_k, b_H^{up}(x_k))$ can be approximated by points from H it follows that $H_s^{low} \neq \emptyset$. Similarly, since $(z_k, b_H^{up}(z_k))$ can be approximated by points from H, $H_s^{up} \neq \emptyset$. Thus, we get a contradiction to Lemma 9.2.3. Therefore $f_1 = f_2$ and $\lim_{x \to x_H^l} b_H^{up}(x)$ exists. □

We denote $b_H^{up}(x_H^l) = \lim_{x \to x_H^l} b_H^{up}(x)$, and similarly for the other three limits. Note that $b_H^{low}(x_H^l) \leq b_H^{up}(x_H^l)$, and both limits belong to $F(x_H^l)$. A similar conclusion hold for the other two limits.

The next lemma shows that the boundaries b_H^{low} and b_H^{up} coincide at the points x_H^l and x_H^r.

Lemma 9.2.6 *For an interior hole* H *of* $F \in CBV[a, b]$

$$b_H^{low}(x_H^l) = b_H^{up}(x_H^l) \quad and \quad b_H^{low}(x_H^r) = b_H^{up}(x_H^r).$$

Proof We prove the lemma for x_H^l. The proof for x_H^r is similar. Assume to the contrary, that $b_H^{low}(x_H^l) < b_H^{up}(x_H^l)$. Let $\tilde{y} \in (b_H^{low}(x_H^l), b_H^{up}(x_H^l))$, clearly $\tilde{y} \in \mathrm{co}(F(x_H^l))$. We now show that $\tilde{y} \in F(x_H^l)$, which implies in view of Lemma 9.2.2 that $p = (x_H^l, \tilde{y})$ is a singular PCT of F in contradiction to the continuity of F.

We exclude the possibility $p \in H$, since H is an open set and since there are no points (x, y) in H with $x < x_H^l$. By similar reasoning $p \notin \widetilde{H} \neq H$, for any $\widetilde{H} \in \mathcal{H}(F)$. Thus $p \in Graph(F)$ which completes the proof. $\square$

Note that for a boundary hole H with $\Delta_H \neq [a, b]$, the statement of Lemma 9.2.6 holds at the end point of Δ_H, which is in (a, b).

As a direct consequence of Lemmas 9.2.4, 9.2.5 and 9.2.6 we get

Theorem 9.2.7 *For $F \in CBV[a, b]$ and $H \in \mathcal{H}(F)$, the functions $b_H^{low}(x)$, $b_H^{up}(x)$ are continuous on $\Delta_H = [x_H^l, x_H^r]$. These two functions coincide at x_H^l, x_H^r for an interior hole, while for a boundary hole they coincide at x_H^l if $x_H^l > a$, or at x_H^r if $x_H^r < b$.*

It is easy to verify that for an interior hole $H \in \mathcal{H}(F)$ of $F \in CBV[a, b]$ the points $(x_H^l, b_H^{low}(x_H^l))$ and $(x_H^r, b_H^{low}(x_H^r))$ are PCTs. For a boundary hole H with $\Delta_H \neq [a, b]$, a similar statement holds at the end point of Δ_H which is inside (a, b).

9.3. Regularity Properties of the Boundaries

In this section we prove that the regularity properties of $F \in CBV[a, b]$ are "inherited" by its boundaries. First we show it for f^{low} and f^{up}.

Theorem 9.3.1 *Let $F : [a, b] \to K(\mathbb{R})$ be a continuous SVF. Then*

$$\omega_{[a,b]}(f^{low}, \delta) \leq \omega_{[a,b]}(F, \delta).$$

Moreover if $F \in CBV[a, b]$, then $V_a^b(f^{low}) \leq V_a^b(F)$.
Similar results are valid for f^{up}.

Proof Let f be either f^{low} or f^{up}. Consider $|f(x) - f(z)|$ with $|x - z| \leq \delta$ and $\delta > 0$. Then from (9.5) and from

$$\text{haus}(F(x), F(z)) \leq \omega_{[a,b]}(F, |x - z|)$$

one easily obtains the first claim of the theorem.

To prove the second claim consider a partition χ. By the definition of variation and by (9.5) we get

$$V(f, \chi) \leq V(F, \chi),$$

which leads to $V_a^b(f) \leq V_a^b(F)$. $\square$

Next we investigate the regularity of $b_H^{low}(x), b_H^{up}(x)$ away from x_H^l, x_H^r. For $\varepsilon > 0$ small enough denote

$$\Delta_H^\varepsilon = [x_H^l + \varepsilon, x_H^r - \varepsilon].$$

Lemma 9.3.2 *Let* $F \in CBV[a, b]$, $H \in \mathcal{H}(F)$. *For* $\varepsilon \in (0, (x_H^r - x_H^l)/2)$ *there exists* $\delta = \delta_{H,\varepsilon} > 0$ *such that for any* $x, z \in \Delta_H^\varepsilon$, *satisfying* $|x - z| \le \delta$,

$$\max\{|b_H^{low}(x) - b_H^{low}(z)|, |b_H^{up}(x) - b_H^{up}(z)|\} \le \mathrm{haus}(F(x), F(z)). \quad (9.14)$$

Moreover,

$$\max\{\omega_{\Delta_H^\varepsilon}(b_H^{low}, \delta), \omega_{\Delta_H^\varepsilon}(b_H^{up}, \delta)\} \le \omega_{[a,b]}(F, \delta). \quad (9.15)$$

Proof Let $\min\{|b_H^{low}(x) - b_H^{up}(x)| : x \in \Delta_H^\varepsilon\} = \gamma > 0$. By the uniform continuity of b_H^{low} and b_H^{up} on Δ_H^ε there exists $\delta > 0$ such that for any $x, z \in \Delta_H^\varepsilon$ satisfying $|z - x| \le \delta$

$$b_H^{up}(z) - b_H^{low}(x) \ge \gamma/2. \quad (9.16)$$

We prove the lemma for b_H^{low}. The proof for b_h^{up} is similar.

Let $|x - z| \le \delta$ and assume without loss of generality that $b_H^{low}(x) \ge b_H^{low}(z)$. We observe that either

$$\Pi_{F(z)}(b_H^{low}(x)) \ni b_H^{low}(z) \quad (9.17)$$

or

$$\min\{y : y \in \Pi_{F(z)}(b_H^{low}(x))\} \ge b_H^{up}(z). \quad (9.18)$$

By definition of the Hausdorff metric

$$\mathrm{dist}(b_H^{low}(x), F(z)) \le \mathrm{haus}(F(z), F(x)). \quad (9.19)$$

In case (9.17) holds we get

$$|b_H^{low}(z) - b_H^{low}(x)| \le \mathrm{haus}(F(z), F(x)),$$

which is the first claim (9.14).

The case (9.18) is impossible, since we get from (9.19), (9.18) and (9.16)

$$\mathrm{haus}(F(z), F(x)) \ge \mathrm{dist}(b_H^{low}(x), F(z)) \ge b_H^{up}(z) - b_H^{low}(x) \ge \gamma/2,$$

in contradiction to the continuity of F.

The second claim of the lemma (9.15) follows by taking the supremum of (9.14) over $|x - z| \leq \delta$.   □

In the closed interval Δ_H the regularity result for b_H^{low}, b_H^{up} is weaker than (9.15).

Theorem 9.3.3 *Let* $F \in CBV[a, b]$. *Then for any* $H \in \mathcal{H}(F)$ *and* $\delta > 0$

$$\max\{\omega_{\Delta_H}(b_H^{low}, \delta), \omega_{\Delta_H}(b_H^{up}, \delta)\} \leq \omega_{[a,b]}(v_F, \delta), \qquad (9.20)$$

with v_F *defined as in* (1.3). *Moreover,*

$$\max\{V_{x_H^l}^{x_H^r}(b_H^{low}), V_{x_H^l}^{x_H^r}(b_H^{up})\} \leq V_{x_H^l}^{x_H^r}(F). \qquad (9.21)$$

Proof Let f be either b_H^{up} or b_H^{low}. By Lemma 9.3.2 for any $\varepsilon > 0$ there exists $\delta_{H,\varepsilon} > 0$ such that for any partition $\chi_H^\varepsilon = \{x_0, x_1, \ldots, x_N\}$ of Δ_H^ε with $x_0 = x_H^l + \varepsilon$ and $x_N = x_H^r - \varepsilon$, satisfying $|\chi_H^\varepsilon| < \delta_{H,\varepsilon}$, we get for $i = 0, 1, \ldots, N - 1$,

$$|f(x) - f(z)| \leq \text{haus}(F(x), F(z)), \quad x, z \in [x_i, x_{i+1}] \qquad (9.22)$$

For $x, z \in \Delta_H^\varepsilon$ with $x \in [x_i, x_{i+1}]$, $z \in [x_j, x_{j+1}]$, $0 \leq i < j \leq N - 1$, it follows from the triangle inequality and (9.22) that

$$|f(x) - f(z)| \leq |f(x) - f(x_{i+1})| + \sum_{l=i+1}^{j-1} |f(x_{l+1}) - f(x_l)| + |f(z) - f(x_j)|$$

$$\leq \text{haus}(F(x), F(x_{i+1})) + \sum_{l=i+1}^{j-1} \text{haus}(F(x_l), F(x_{l+1}))$$

$$+ \text{haus}(F(z), F(x_j)). \qquad (9.23)$$

Now using the definition of the variation of F and (1.3) we get for $x, z \in \Delta_H^\varepsilon$, $|x - z| \leq \delta$

$$|f(x) - f(z)| \leq V_x^z(F) = v_F(z) - v_F(x) \leq \omega_{[a,b]}(v_F, \delta). \qquad (9.24)$$

Taking the supremum over $|x - z| \leq \delta$, $x, z \in \Delta_H^\varepsilon$ we obtain

$$\omega_{\Delta_H^\varepsilon}(f, \delta) \leq \omega_{[a,b]}(v_F, \delta)$$

which holds for any $\varepsilon > 0$. In view of Theorem 9.2.7, the last inequality leads to the first claim of the theorem.

By the first inequality in (9.24) and the additivity property of the variation, we get (9.21), but with $V_{x_H^l}^{x_H^r}$ replaced by $V_{x_H^l+\varepsilon}^{x_H^r-\varepsilon}$. Taking $\varepsilon \to 0$ we obtain (9.21). □

For Lipschitz continuous SVFs we have a stronger result.

Theorem 9.3.4 *If* $F \in Lip([a,b], L)$, *then for any* $f \in \partial F$

$$f \in Lip(D_f, L), \tag{9.25}$$

with D_f *the domain of definition of* f.

Proof Let f be either b_H^{up} or b_H^{low} for some $H \in \mathcal{H}(F)$. Following the first steps in the proof of Theorem 9.3.3 we get, instead of (9.22),

$$|f(x) - f(z)| \leq \text{haus}(F(x), F(z)) \leq L|x - z|, \quad x, z \in [x_i, x_{i+1}] \subset \Delta_H^\varepsilon$$

and, instead of (9.23),

$$|f(x) - f(z)| \leq L|x_{i+1} - x| + \sum_{l=i+1}^{j-1} L|x_{l+1} - x_l| + L|y - x_j| = L|z - x|,$$

for any $x, z \in \Delta_H^\varepsilon$ and any $\varepsilon > 0$ small enough. Taking $\varepsilon \to 0$ we get in view of Theorem 9.2.7 that (9.25) holds for this f.

In case f is either f^{up} or f^{low} the proof is even simpler, based on (9.5) instead of (9.22). □

The theorems in this section guarantee the regularity properties of the boundaries of a CBV multifunction, which are applied in the next chapters to the construction of adapted approximation operators and to their analysis.

From now on we consider continuous SVFs of bounded variation with a finite number of holes in their graphs. The last assumption facilitates the derivation of multi-segmental representations in the next chapter.

Chapter 10

Multi-Segmental and Topological Representations

In this chapter we introduce and investigate multi-segmental representations of CBV multifunctions with a finite number of holes in their
graphs. In such a representation the multifunction is given as a finite
union of segment functions, namely SVFs which have segments as their
images. The segment functions in a multi-segmental representation (MSR)
intersect at most at their boundaries, which constitute the boundaries
of the MSR. Among all continuous MSRs we single out the topological
MSRs, having boundaries with regularity properties determined by those
of the multifunction. We propose a special construction of topological
MSRs in which the number of segment functions is minimal. Also, we
determine conditions on the SVF guaranteeing that such a representation
is unique. From a topological MSR of a multifunction F we construct a
complete representation of F in terms of topological selections.

In the next chapter we use these representations to adapt approximation operators.

10.1. Multi-Segmental Representations (MSRs)

We denote by $\mathcal{F}[a, b]$ the collection of CBV multifunctions on $[a, b]$ with
finite number of holes in their graphs. All SVFs in the rest of the book are
from $\mathcal{F}[a, b]$.

First we notice that any image $F(x)$ of $F \in \mathcal{F}[a, b]$ for $x \in [a, b]$ consists of a finite number of disjoint, closed intervals of R, termed segments. Namely,

$$F(x) = \bigcup_{n=1}^{N(x)} [a_n(x), b_n(x)], \quad x \in [a, b] \tag{10.1}$$

with $a_n(x) \leq b_n(x)$, $n = 1, \ldots, N(x)$, $b_n(x) < a_{n+1}(x)$, $n = 1, \ldots, N(x)-1$. Clearly, $\{a_n(x), b_n(x)\}_{n=1}^{N(x)}$ are points on the boundary of $Graph(F)$.

If $N(x) \equiv 1$ then F is a **segment function** and can be represented as

$$F(x) = [f^{low}(x), f^{up}(x)], \quad f^{low}(x) \leq f^{up}(x), \; x \in [a, b], \tag{10.2}$$

with $f^{low}(x) = a_1(x)$ and $f^{up}(x) = b_1(x)$.

In this case F is a convex-valued function. It is easy to prove (e.g. by Theorem 9.2.1 and by (2.1)) that the segment function F is continuous iff f^{low} and f^{up} are continuous.

Definition 10.1.1 A multifunction $F \in \mathcal{F}[a, b]$ has a **multi-segmental representation** (MS-representation, MSR), if there is a natural number N such that

$$F(x) = \bigcup_{n=1}^{N} F_n(x) = \bigcup_{n=1}^{N} [f_n^{low}(x), f_n^{up}(x)], \quad x \in [a, b], \tag{10.3}$$

where $F_n = [f_n^{low}, f_n^{up}]$, $n = 1, \ldots, N$ are segment functions, and where for $x \in [a, b]$

$$f_1^{low}(x) \leq f_1^{up}(x) \leq f_2^{low}(x) \leq f_2^{up}(x) \leq \cdots$$
$$\leq f_{N-1}^{up}(x) \leq f_N^{low}(x) \leq f_N^{up}(x). \tag{10.4}$$

By definition, the functions in (10.4) are selections of F. We denote such a multi-segmental representation by

$$\mathcal{R} = \{F_n, \; n = 1, \ldots, N\}. \tag{10.5}$$

$\mathcal{R}$ is determined by the family of selections

$$\mathcal{B}(\mathcal{R}, F) = \{f_n^{low}, f_n^{up}, \; n = 1, \ldots, N\}. \tag{10.6}$$

We call each selection in (10.6) an **MS-boundary**. An example of such a representation is shown in Fig. 10.1.2(a). In general, the MS-boundaries may be quite arbitrary. Yet there is a class of SVFs which have MS-representations determined by their boundaries.

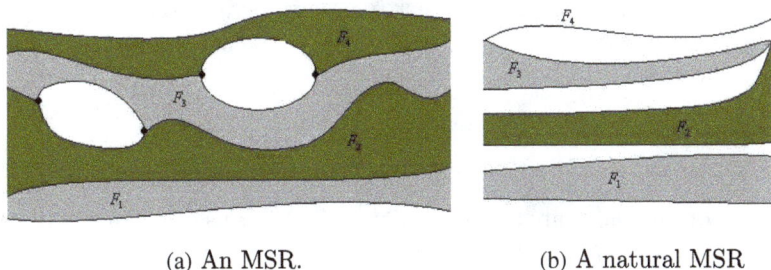

(a) An MSR.   (b) A natural MSR

Figure 10.1.2   An MSR and a natural MSR.

Definition 10.1.3 An MS-representation $\mathcal{R}$ of a multifunction $F \in \mathcal{F}[a, b]$ is called **natural** if $\mathcal{B}(\mathcal{R}, F) = \partial F$ and $\{x : F_i(x) \cap F_{i+1}(x) \neq \emptyset\} \subset \{a, b\}$ for any $1 \leq i \leq N - 1$. The boundaries $\mathcal{B}(\mathcal{R}, F)$ are called **natural MS-boundaries**. Also any $f \in \partial F$ with $D_f = [a, b]$ is called a natural MS-boundary.

Clearly, the natural MS-representation is unique. The graph of a multifunction F with its natural MSR is shown in Fig. 10.1.2(b).

The following observation is central in the definition of topological MSRs, investigated in Section 10.2.

Remark 10.1.4 Any multifunction $F \in \mathcal{F}[a, b]$ determines a **natural partition** of $[a, b]$, $\chi_F = \{x_0, \ldots, x_M\}$, consisting of the distinct points among the points $\{a, b\} \cup \{x_H^l, x_H^r : H \in \mathcal{H}(F)\}$. For each $\Delta_i = [x_i, x_{i+1}]$, $i = 0, \ldots, M - 1$, we define

$$\mathcal{H}_i(F) = \{H : H \in \mathcal{H}(F), \Delta_i \subseteq \Delta_H\},$$

and observe that

$$\partial F|_{\Delta_i} = \{f^{up}|_{\Delta_i}, f^{low}|_{\Delta_i}\} \bigcup \{b_H^{up}|_{\Delta_i}, b_H^{low}|_{\Delta_i} : H \in \mathcal{H}_i(F)\}.$$

It is easy to see that

$$F(x) = [f^{low}(x), f^{up}(x)] \Big\backslash \bigcup_{H \in \mathcal{H}_i(F)} (b_H^{low}(x), b_H^{up}(x)), \quad x \in \Delta_i \qquad (10.7)$$

which is equivalent to

$$F|_{\Delta_i}(x) = \bigcup_{n=1}^{N_i} F_n^i(x), \quad x \in \Delta_i, \qquad (10.8)$$

where F_n^i, $n = 1, \ldots, N_i$ are segment functions, defined on Δ_i. We denote the boundaries of F_n^i by $(f^i)_n^{low}, (f^i)_n^{up} \in \partial F|_{\Delta_i}$. Thus $F_n^i = [(f^i)_n^{low}, (f^i)_n^{up}]$ and

$$(f^i)_n^{low}(x) \le (f^i)_n^{up}(x) \le (f^i)_{n+1}^{low}(x), \quad x \in \Delta_i.$$

Since by the definition of Δ_i, by (10.7) and by Theorem 9.2.7 equality can occur only at the endpoints of Δ_i, this is a natural MS-representation of $F|_{\Delta_i}$. Thus any $F \in \mathcal{F}[a,b]$ has a unique **piecewise natural MS-representation**.

To illustrate Remark 10.1.4, consider F with $Graph(F)$ shown in Fig. 10.2.5. For this F the natural partition is $\chi_F = \{a, x_H^l, x_H^r, b\}$, $\Delta_0 = [a, x_H^l]$, $\Delta_1 = [x_H^l, x_H^r]$, $\Delta_2 = [x_H^r, b]$, and (10.8) holds with $N_0 = 2$, $N_1 = 2$, $N_2 = 1$. The natural MS-boundaries on Δ_0 are

$$(f^0)_1^{low} = f^{low}|_{\Delta_0}, \quad (f^0)_1^{up} = b_{\widetilde{H}}^{low}, \quad (f^0)_2^{low} = b_{\widetilde{H}}^{up}, \quad (f^0)_2^{up} = f^{up}|_{\Delta_0}.$$

Similarly, the natural MS-boundaries on Δ_1 and Δ_2 can be easily defined from Fig. 10.2.5.

From now on we consider only multi-segmental representations with continuous MS-boundaries and call them **continuous MS-representations**.

Remark 10.1.5

(i) The segment functions in a continuous MSR are continuous.
(ii) A multifunction with a continuous MSR is continuous.
(iii) A multifunction with a discontinuous MSR is not necessarily discontinuous.

Figure 10.1.7 illustrates a discontinuous MSR of a continuous F.

It is clear that for any multifunction $F \in \mathcal{F}[a,b]$ and any of its MS-representations $\mathcal{R}$ of the form (10.5), the following holds

$$\bigcup_{f \in \partial F} Graph(f) \subset \bigcup_{b \in \mathcal{B}(\mathcal{R},F)} Graph(b).$$

For continuous MSRs a stronger result holds, namely that every boundary function of F must be extended continuously to an MS-boundary.

Lemma 10.1.6 *Let $F \in \mathcal{F}[a,b]$ and let $\mathcal{R}$ be a continuous MSR of F. Then for any $f \in \partial F$ there exists $b \in \mathcal{B}(\mathcal{R}, F)$ such that $b(x) = f(x)$, for $x \in D_f$.*

Proof If $f = f^{low}$ or $f = f^{up}$, then the claim of the lemma is trivial. Consider $f = b_H^{up}$ for some $H \in \mathcal{H}(F)$. First we show that any point of $Graph(f)$ belongs to the graph of some MS-boundary. Let $y = f(x)$, for a fixed $x \in \Delta_H$. Obviously $(x,y) \in Graph(F_i)$, where F_i is some segment function in (10.3). It is clear that $(x,y) \in \partial Graph(F_i)$, since $(x,y) \in \partial Graph(F)$ and $Graph(F_i) \subset Graph(F)$. Denote by s_x a selection from $\mathcal{B}(\mathcal{R}, F)$, satisfying $s_x(x) = b_H^{up}(x)$.

Next we prove that there must be an MS-boundary, which coincides with b_H^{up} on Δ_H. Let

$$S_H^{up} = \{s : s \in \mathcal{B}(\mathcal{R}, F), s(x) \geq b_H^{up}(x), x \in \Delta_H\},$$

and let $s_{min} \in S_H^{up}$ be such that $s_{min}(x) \leq s(x)$, $x \in [a,b]$ for any $s \in S_H^{up}$. Such a selection exists by (10.4).

To see that b_H^{up} coincides on Δ_H with s_{min} assume the contrary, which in view of Lemma 9.2.3 is $s_{min} > b_H^{up}$ on a non-empty subset $\widetilde{\Delta}_H$ of Δ_H. By the continuity of s_{min} and b_H^{up}, one may assume that $\widetilde{\Delta}_H$ contains an interior point $x^* \in (x_H^l, x_H^r)$. Thus there exists $s_{x^*} \in \mathcal{B}(\mathcal{R}, F)$ such that $s_{x^*}(x^*) = b_H^{up}(x^*)$. It follows from Lemma 9.2.3 that $s_{x^*} \in S_H^{up}$. The above observation and assumption lead to

$$b_H^{up}(x^*) < s_{min}(x^*) \leq s_{x^*}(x^*) = b_H^{up}(x^*),$$

which is a contradiction. Thus $b_H^{up} = s_{min}$ on Δ_H. A similar proof applies for $f = b_H^{low}$. $\square$

Figure 10.1.7 demonstrates that the continuity of the MS-boundaries is essential in Lemma 10.1.6. In that figure the boundaries $f_2^{low}(x)$ and $f_2^{up}(x)$ of the segment function F_2 are discontinuous at x_0. Indeed

$$f_2^{low}(x) = \begin{cases} C_1, & x \in [x_H^l, x_0], \\ b_H^{up}(x), & x \in (x_0, x_H^r], \end{cases}$$

$$f_2^{up}(x) = \begin{cases} b_H^{low}(x), & x \in [x_H^l, x_0], \\ C_2, & x \in (x_0, x_H^r], \end{cases}$$

where $C_1 = b_H^{low}(x_0)$ and $C_2 = b_H^{up}(x_0)$ are constants. Thus there is no $b \in \mathcal{B}(\mathcal{R}, F)$ which coincides either with b_H^{up} or with b_H^{low} on the whole domain Δ_H. Yet each point on the boundary of H belongs to the graph of some $\mathcal{B}(\mathcal{R}, F)$. Indeed $b_H^{up}(x) = f_3^{low}(x)$, $x \in (x_H^l, x_0)$ and $b_H^{up}(x) = f_2^{low}(x)$, $x \in (x_0, x_H^r)$, while $b_H^{low}(x) = f_2^{up}(x)$, $x \in (x_H^l, x_0)$ and $b_H^{low}(x) = f_1^{up}(x)$, $x \in (x_0, x_H^r)$.

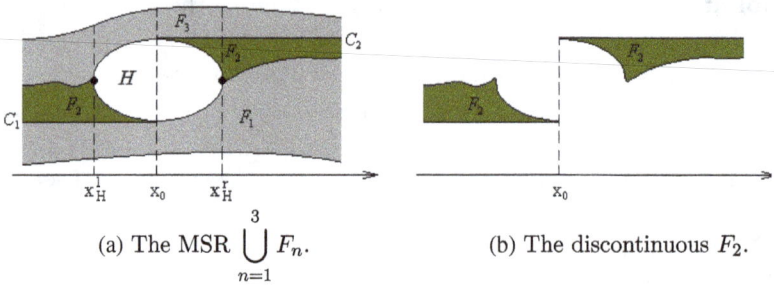

(a) The MSR $\bigcup\limits_{n=1}^{3} F_n$.   (b) The discontinuous F_2.

Figure 10.1.7   A discontinuous MSR of a continuous SVF.

It follows from Lemma 10.1.6 that the boundaries of any continuous MSR, contain a special set of selections termed hereafter significant selections.

Definition 10.1.8 A selection of $F \in \mathcal{F}[a,b]$ is called a **significant selection** if it is continuous on $[a,b]$ and coincides with some $f \in \partial F$ on D_f.

In Fig. 10.1.2(a) all the selections in $\mathcal{B}(\mathcal{R}, F)$ except $f_1^{up} = f_2^{low}$ are significant selections. It should be noted that the number of segment functions in an MSR can be reduced by deleting the non-significant selections from the MS-boundaries. In the next section we construct continuous MS-representations with MS-boundaries consisting only of significant selections.

10.2.   Topological MSRs

One can construct various multi-segmental representations of F by different selections. Moreover, the number of segment functions in (10.3) may be arbitrarily large. Lemma 10.1.6 shows that significant selections must participate in any continuous MSR of $F \in \mathcal{F}[a,b]$.

We propose here an MS-representation with MS-boundaries which are special significant selections. These selections inherit the behavior of the boundaries of F and are termed topological. We call the resulting representation "topological MSR".

From this point on we use the notation of Remark 10.1.4.

Definition 10.2.1 A significant selection s of $F \in \mathcal{F}[a,b]$ is called a **significant topological selection** (ST-selection) if for each Δ_i, where s does not coincide with $f \in \partial F$, there exist $n \in \{1, \ldots, N_i\}$ and $\lambda_n^i \in [0,1]$

such that

$$s|_{\Delta_i} = \lambda_n^i (f^i)_n^{low} + (1 - \lambda_n^i)(f^i)_n^{up},$$

where the functions $(f^i)_n^{low}$, $(f^i)_n^{up}$ are determined by the piecewise natural MS-representation (10.8).

Clearly $f^{up}, f^{low} \in \partial F$ and all the natural MS-boundaries are ST-selections of F. An example of two ST-selections is given in Fig. 10.2.5. The two selections coincide on $\Delta_0 \cup \Delta_2$, while on Δ_1 one is equal to b_H^{up} and the other to b_H^{low}.

Definition 10.2.2 An MS-representation with MS-boundaries which are ST-selections is called a **topological MS-representation** (TMSR).

In the following we prove the existence of a TMSR and give conditions guaranteeing its uniqueness.

10.2.1. *Existence of a topological MSR*

Our proof of the existence of a TMSR is constructive. The construction uses only a special subset of ST-selections.

Definition 10.2.3 For each $H \in \mathcal{H}(F)$ we define a **pair of ST-selections** t_H^{up}, t_H^{low} by

$$t_H^{up} = b_H^{up}, \quad t_H^{low} = b_H^{low} \text{ on } \Delta_H$$
$$t_H^{up} = t_H^{low} \quad \text{on } [a, b] \setminus \Delta_H.$$

Note that in general a hole $H \in \mathcal{H}(F)$ may have more than one pair of ST-selections (see discussion in subsection 10.2.2). Figure 10.2.5 illustrates a hole H with a unique pair of ST-selections.

Our construction of a TMSR is recursive. Each step starts with a union of multifunctions representing F, and eliminates ("cuts") one hole of one of these multifunctions, replacing it by two SVFs. The cutting of the hole is along one of its pairs of ST-selections. Thus at the end of such a step the number of multifunctions representing F is increased by one, while the total number of holes of these SVFs is decreased by one. The number of steps required for the construction of a TMSR of F by this procedure is $|\mathcal{H}(F)|$ which is finite by assumption. We first describe the idea of the construction on an example.

Example 10.2.4 Consider F with the graph presented in Fig. 10.2.5.

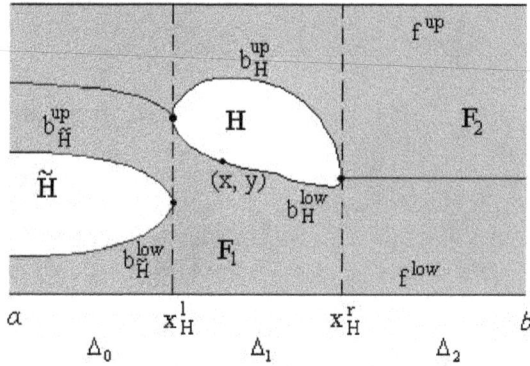

Figure 10.2.5　Eliminating a hole.

To construct our TMSR of F, we define a pair of ST-selections corresponding to the hole H. The first selection t_H^{low} coincides with b_H^{low} on Δ_1, while on Δ_0 and Δ_2 it is defined as a fixed convex combination of the relevant MS-boundaries. More precisely

$$t_H^{low}(z) = b_H^{low}(z), \quad z \in \Delta_1 = [x_H^l, x_H^r],$$

$$t_H^{low}(z) = \lambda_0 b_{\widetilde{H}}^{up}(z) + (1 - \lambda_0) f^{up}(z), \quad z \in \Delta_0 = [a, x_H^l],$$

$$t_H^{low}(z) = \lambda_2 f^{low}(z) + (1 - \lambda_2) f^{up}(z), \quad z \in \Delta_2 = [x_H^r, b],$$

with $\lambda_0 = (f^{up}(x_H^l) - b_H^{up}(x_H^l))/(f^{up}(x_H^l) - b_{\widetilde{H}}^{up}(x_H^l))$, guaranteeing the continuity of t_H^{low} at x_H^l. To guarantee the continuity of t_H^{low} at x_H^r we define $\lambda_2 = (f^{up}(x_H^r) - b_H^{up}(x_H^r))/(f^{up}(x_H^r) - f^{low}(x_H^r))$.

The second selection t_H^{up} coincides with b_H^{up} on Δ_1 and with t_H^{low} on $\Delta_0 \cup \Delta_2$. Thus t_H^{low}, t_H^{up} partition $Graph(F)$ into two subgraphs (as depicted in Fig. 10.2.5), such that

$$F(x) = F_1(x) \bigcup F_2(x),$$

with $f_1^{low}(x) = f^{low}(x)$, $f_1^{up}(x) = t_H^{low}(x)$, $f_2^{low}(x) = t_H^{up}(x)$, $f_2^{up}(x) = f^{up}(x)$ the lower and upper boundaries of $F_1(x)$ and $F_2(x)$ respectively.

Note that F_2 is segmental, but F_1 still has non-convex images. The graph of F_1 has a unique hole $\widetilde{H}$, while H is no longer a hole of F_1 or of F_2. Next, using the same technique, the hole $\widetilde{H}$ can be eliminated by cutting along the pair of ST-selections of $\widetilde{H}$ in $\Delta_1 \cup \Delta_2$. This leads to a subdivision of F_1 into two segment functions. The union of these two multifunctions with F_2 gives a TMSR of the original SVF. It is easy to verify that in this example the same TMSR is obtained when we first eliminate $\widetilde{H}$ and then H.

In the following we describe the construction of a TMSR of $F \in \mathcal{F}[a,b]$ in the form of an algorithm. We use here the notation $t^{up}(F)$ $(t^{low}(F))$ for f^{up} (f^{low}) of F.

Construction procedure.

Given a multifunction $F \in \mathcal{F}[a,b]$

(i) Set $i = 1$, $F_i = F$, $I = 1$.
(ii) **While** $\langle\ i \leq I\ \rangle$

    If $\langle\ |\mathcal{H}(F_i)| > 0\ \rangle$

        (1) Choose any hole $H \in \mathcal{H}(F_i)$.
        (2) Construct t_H^{low}, t_H^{up} of F_i.
        (3) $I = I + 1$.
        (4) $F_I(x) = F_i(x) \cap [t^{low}(F_i)(x), t_H^{low}(x)], \quad x \in [a,b]$.
        (5) $F_i(x) = F_i(x) \cap [t_H^{up}(x), t^{up}(F_i)(x)], \quad x \in [a,b]$.

    Else $i = i + 1$.

    End While
(iii) For any $x \in [a,b]$, $F(x) = \bigcup_{i=1}^{I} F_i(x)$.

The obtained segment functions $\{F_i\}$ in the above procedure have disjoint interiors, but are not ordered in the sense of Definition 10.1.1. This can be corrected by renumbering these multifunctions. Also note that the construction yields an MS-representation with significant selections as MS-boundaries. From the next lemma and remark we conclude that this MSR is topological.

Remark 10.2.6 For any two functions f_1, f_2 defined on an interval Δ, let $g_i = \lambda_i f_1 + (1 - \lambda_i) f_2$, with $\lambda_i \in [0,1]$, $i = 1, 2$. Any function of the form

$$h = \mu g_1 + (1 - \mu) g_2, \quad \mu \in [0,1]$$

is a convex combination of f_1 and f_2.

Lemma 10.2.7 *Let* $G(x) \subset F(x)$ *for* $x \in [a,b]$ *such that* $\mathcal{H}(G) \subset \mathcal{H}(F)$ *and* $t^{up}(G)$, $t^{low}(G)$ *are ST-selections of* F. *Then any ST-selection of the multifunction* G *is a ST-selection of* F.

Proof We use here the notation of Remark 10.1.4. Clearly by the assumptions of the lemma, χ_G is a subset of χ_F. Let s be an ST-selection of G. By Definitions 10.1.8 and 10.2.1, if s coincides with $t^{up}(G)$ or $t^{low}(G)$ then by assumption it is an ST-selection of F. Otherwise if $s = t_H^{up}$ or t_H^{low} for some $H \in \mathcal{H}(G)$, then on any interval $\Delta \neq \Delta_H$ specified by χ_G

$$s|_{\Delta} = \lambda g_n^{low} + (1 - \lambda)g_n^{up}, \tag{10.9}$$

for some $\lambda \in [0, 1]$ and $g_n^{low}, g_n^{up} \in \partial G|_{\Delta}$. But since $\mathcal{H}(G) \subset \mathcal{H}(F)$ and in view of the assumption on $t^{low}(G), t^{up}(G)$ and Remark 10.2.6, s is an ST-selection of F on each interval Δ determined by χ_F, and therefore on $[a, b]$. □

The advantage of a TMSR is that its boundaries inherit the regularity properties of F. This follows from Theorems 9.3.1, 9.3.3, 9.3.4 and Definition 10.2.1. Thus we have

Corollary 10.2.8 *Let $F \in \mathcal{F}[a, b]$ and let $\mathcal{R}$ be a TMSR of F. Then for any $f \in \mathcal{B}(\mathcal{R}, F)$*

$$\omega_{[a,b]}(f, \delta) \leq \omega_{[a,b]}(v_F, \delta), \ \delta > 0 \quad and \quad V_a^b(f) \leq V_a^b(F). \tag{10.10}$$

Moreover, if $F \in Lip([a, b], L)$ then $f \in Lip([a, b], L)$.

10.2.2. *Conditions for uniqueness of a TMSR*

A topological MSR is not necessarily unique. Figure 10.2.9(a) illustrates the graph of a multifunction F with two possible TMSRs of the form $F = F_1 \cup F_2 \cup F_3$. Here F_2 is not defined uniquely on Δ_{H_1}, as is demonstrated by Figs. 10.2.9(b), (c).

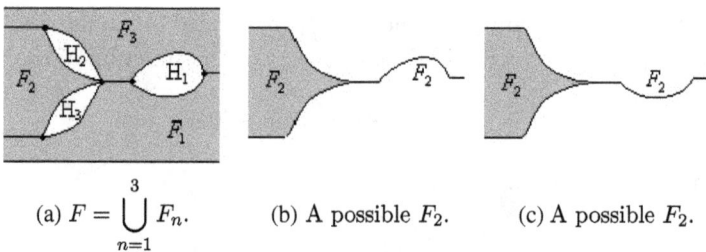

(a) $F = \bigcup_{n=1}^{3} F_n$.   (b) A possible F_2.   (c) A possible F_2.

Figure 10.2.9 Two TMSRs of F.

Moreover, there exist other TMSRs since, on $\Delta_{H_2} = \Delta_{H_3}$ the ST-selections $t_{H_1}^{low}$, $t_{H_1}^{up}$ are not uniquely defined.

The main source of non-uniqueness in our construction of a TMSR is the non-uniqueness of a pair of ST-selections of a hole. The pair t_H^{low}, t_H^{up} for $H \in \mathcal{H}(F)$ is unique if on $[a,b] \backslash \Delta_H$, the graph of $t_H^{low} = t_H^{up}$ is contained in the interior of $Graph(F)$. Non-uniqueness can occur when a pair of ST-selections t_H^{low}, t_H^{up} of a hole H passes through a point of change of topology associated with another hole $\widetilde{H}$, as in the examples in Fig. 10.2.9 and 10.2.10. Note that this is a non-generic situation.

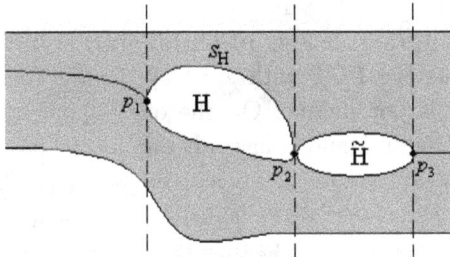

Figure 10.2.10 *F* with non-unique pairs of ST-selections.

To eliminate this source of non-uniqueness we use in the construction another type of pairs of ST-selections.

Definition 10.2.11 For each $H \in \mathcal{H}(F)$ we define a **special ST-pair** $(\theta_H^{up}, \theta_H^{low})$ by:

$$\theta_H^{up} = b_H^{up}, \quad \theta_H^{low} = b_H^{low} \quad \text{on } \Delta_H.$$

On every $\Delta_i \neq \Delta_H$ determined by χ_F, $\theta_H^{up} = \theta_H^{low}$ except if the graphs of θ_H^{up} and θ_H^{low} pass through a PCT associated with $\widetilde{H} \neq H$. Then $\theta_H^{up} = b_{\widetilde{H}}^{up}$ and $\theta_H^{low} = b_{\widetilde{H}}^{low}$ on $\Delta_{\widetilde{H}}$.

Remark 10.2.12 Note that the pair $(\theta_H^{up}, \theta_H^{low})$ coincides with the pair (t_H^{up}, t_H^{low}) whenever (t_H^{up}, t_H^{low}) is uniquely defined, otherwise it is a pair of ST-selections associated with more than one hole (as in Fig. 10.2.10).

Still, in some cases there is ambiguity in the construction of $(\theta_H^{up}, \theta_H^{low})$ (see Fig. 10.2.9). The uniqueness of the special topological pairs can be guaranteed if all the points of change of topology of F are regular in the sense of the following definition.

Definition 10.2.13 In the notation of Remark 10.1.4, a PCT (x, y) is called **regular** if for all i and n such that $x \in \Delta_i$, $y \in F_n^i(x)$,

$$\lim_{z \to x} \mu(F_n^i(z)) > 0,$$

where μ is the Lebesgue measure of $\mathbb{R}$.

It follows from the above definition that any regular point of change of topology (x, y) satisfies $y \neq f^{low}(x), f^{up}(x)$, and is associated with at most two holes. In the case that two holes H and $\widetilde{H}$ are associated with (x, y), then $\Delta_{\widetilde{H}} \cap \Delta_H = \{x\}$ (see Fig. 10.2.10).

It is easy to see that in Fig. 10.2.9 $\mu(F_2(z)) \to 0$ as z tends to $x_{H_3}^r$ from the left. Thus the PCT with $x = x_{H_3}^r = x_{H_2}^r$ is not regular, and indeed $(\theta_{H_1}^{up}, \theta_{H_1}^{low})$ is not unique. On the other hand in Fig. 10.2.10 the PCTs p_1, p_2, p_3 are regular and F has only one special pair of ST-selections, since the pair $(\theta_H^{up}, \theta_H^{low})$ coincides with the pair $(\theta_{\widetilde{H}}^{up}, \theta_{\widetilde{H}}^{low})$. The TMSR $F = F_1 \cup F_2$ with $F_1 = [f^{low}, \theta_H^{low}]$, $F_2 = [\theta_H^{up}, f^{up}]$ is minimal in the sense that there is no function among the segment functions that can be removed from the representation.

In general we have

Lemma 10.2.14 *For F with only regular PCTs, each $H \in \mathcal{H}(F)$ determines a unique special ST-pair $(\theta_H^{up}, \theta_H^{low})$.*

Proof We show the uniqueness of θ_H^{up} by induction. The proof for θ_H^{low} is similar. In the proof we use the notation of Remark 10.1.4.

Obviously $\theta_H^{up} = b_H^{up}$ on Δ_H, and θ_H^{up} is uniquely defined on any interval containing Δ_H but no other $\Delta_{\widetilde{H}}$ for $\widetilde{H} \neq H$. Suppose that θ_H^{up} is defined uniquely on $[x_j, x_{j+k}] \supset \Delta_H$, $x_j, x_{j+k} \in \chi_F$. First we consider the case $j > 0$. If $(x_j, \theta_H^{up}(x_j))$ is a regular PCT associated with some $\widetilde{H} \in \mathcal{H}(F)$, $\widetilde{H} \neq H$ then $x_j = x_{\widetilde{H}}^r$ and by definition $\theta_H^{up} = b_{\widetilde{H}}^{up}$ on $\Delta_{\widetilde{H}}$. Otherwise θ_H^{up} is determined uniquely on $[x_{j-1}, x_j]$ by Definition 10.2.1. Thus θ_H^{up} is defined uniquely on $[x_{j-1}, x_{j+k}]$. In case $j = 0$, and $[x_0, x_k] \neq [a, b]$ then $k < M$, and similar arguments guarantee that θ_H^{up} is uniquely defined on $[x_0, x_{k+1}]$.   □

Finally, we conclude from the last lemma, Remark 10.2.12 and Definition 10.2.11 that

Corollary 10.2.15 *Let $H, \widetilde{H}$ be two distinct holes of a multifunction F which has only regular PCTs. If the pair $(\theta_H^{up}, \theta_H^{low})$ is different from*

$(\theta_{\tilde{H}}^{up}, \theta_{\tilde{H}}^{low})$ *then there exists* $\varepsilon \in \{-1, 1\}$ *such that for all* $x \in [a, b]$,

$$\varepsilon \theta_H^{low}(x) < \varepsilon \theta_{\tilde{H}}^{low}(x) \quad and \quad \varepsilon \theta_H^{up}(x) < \varepsilon \theta_{\tilde{H}}^{up}(x).$$

Definition 10.2.16 A TMSR determined by the distinct special ST-pairs among

$$\{(\theta_H^{up}, \theta_H^{low}) : H \in \mathcal{H}(F)\} \tag{10.11}$$

is termed **least TMSR** and is denoted by $\mathcal{R}^*$.

Using Corollary 10.2.15, we show in the next theorem that a least TMSR has a minimal number of segment functions among all TMSRs of F.

Theorem 10.2.17 *If all PCTs of F are regular, then F has a unique least TMSR.*

Proof By Lemma 10.2.14 the set (10.11) is uniquely determined. Now by Corollary 10.2.15 all distinct pairs among the pairs in (10.11) can be ordered. Denoting the number of these pairs by $m \leq |\mathcal{H}(F)|$, we obtain

$$f^{low} \leq \theta_{H_1}^{low} \leq \theta_{H_1}^{up} \leq \theta_{H_2}^{low} \leq \cdots \leq \theta_{H_m}^{low} \leq \theta_{H_m}^{up} \leq f^{up},$$

where H_i is a hole associated with the i-th pair in the sequence of distinct special ST-pairs.

Define the segment functions $F_1 = [f^{low}, \theta_{H_1}^{low}]$, $F_{m+1} = [\theta_{H_m}^{up}, f^{up}]$ and $F_i = [\theta_{H_{i-1}}^{up}, \theta_{H_i}^{low}]$, $i = 2, \ldots, m$. Now we show that $F = \bigcup_{i=1}^{m+1} F_i$. Observe that for a given $x \in [a, b]$

$$\left(\bigcup_{i=1}^{m+1} F_i\right)(x) = \bigcup_{i=1}^{m+1} \left[\theta_{H_{i-1}}^{up}(x), \theta_{H_i}^{low}(x)\right],$$

where $\theta_{H_0}^{up}(x) = f^{low}(x)$ and $\theta_{H_{m+1}}^{low}(x) = f^{up}(x)$. Denoting

$$\mathcal{H}_x(F) = \{H \in \mathcal{H}(F) : x \in (x_H^l, x_H^r)\},$$

we obtain,

$$\left(\bigcup_{i=1}^{m+1} F_i\right)(x) = [f^{low}(x), f^{up}(x)] \setminus \bigcup_{H \in \mathcal{H}_x(F)} (b_H^{low}(x), b_H^{up}(x)), \quad x \in [a, b].$$

Obviously $F(x)$ coincides with the right-hand side of the above equality. Hence $F_1, \ldots, F_{m+1}$ determine a TMSR of F.

It remains to show that any other TMSR consists of more than $m + 1$ segment functions. Since each pair $(\theta_H^{up}, \theta_H^{low})$ is associated with the maximal possible number of holes, any other TMSR of F consists of at least $m + 1$ segment functions, and if the number is $m + 1$ then it is the TMSR defined by the distinct special topological pairs in (10.11). □

In the next chapter approximation operators are adapted to SVFs, based on least TMSRs.

10.3.   Representation by Topological Selections

Any TMSR of a multifunction F induces a complete representation of F by topological selections with regularity properties "inherited" from F. We call both representations **topological representations**. Both are the basis for the adaptations of approximation operators studied in the next chapter. While the adaptation based on a TMSR is limited to positive operators, the one based on the complete representation is not.

Definition 10.2.1 introduces the notion of significant topological selections which are the MS-boundaries of a TMSR. Here we extend this notion to topological selections (which are not necessarily significant).

Definition 10.3.1 Let $\mathcal{R} = \{F_n = [f_n^{low}, f_n^{up}], n = 1, \ldots, N\}$ be a TMSR of $F \in \mathcal{F}[a, b]$, namely

$$F(x) = \bigcup_{n=1}^{N} F_n(x) = \bigcup_{n=1}^{N} [f_n^{low}(x), f_n^{up}(x)].$$

We call a continuous selection $s(x)$ of F **topological** if for some $n \in \{1, 2, \ldots, N\}$ and some $\lambda \in [0, 1]$

$$s(x) = s_n^\lambda(x) = \lambda f_n^{low}(x) + (1 - \lambda) f_n^{up}(x), \quad x \in [a, b]. \tag{10.12}$$

We use the notation of Definition 10.3.1, in the rest of this section.

Several properties of topological selections are derived in the following lemmas.

Lemma 10.3.2 *Through any point* $(x_0, y_0) \in Graph(F)$, *there exists a topological selection* $s_n^\lambda(x)$, *with* n *and* λ *depending on* (x_0, y_0).

Proof Since $(x_0, y_0) \in Graph(F)$, there is a segment function F_n such that $(x_0, y_0) \in Graph(F_n)$. Therefore there exists $\lambda \in [0, 1]$ such that

$y_0 = \lambda f_n^{low}(x_0) + (1 - \lambda) f_n^{up}(x_0)$. Thus the topological selection through (x_0, y_0) is $s_n^\lambda(x)$ given by (10.12). □

It follows directly from Corollary 10.2.8 that

Lemma 10.3.3 *The topological selection s_n^λ of the form* (10.12) *satisfies*

$$\omega_{[a,b]}(s_n^\lambda, \delta) \le \omega_{[a,b]}(v_F, \delta), \ \delta > 0, \quad and \quad V_a^b(s_n^\lambda) \le V_a^b(F). \qquad (10.13)$$

Furthermore, if $F \in Lip([a, b], L)$, then also $s_n^\lambda \in Lip([a, b], L)$.

The next lemma follows from the order (10.4) of the functions in $\mathcal{B}(\mathcal{R}, F)$ and from Definition 10.3.1.

Lemma 10.3.4 *For any $x \in [a, b]$, $s_n^{\lambda_1}(x) \le s_m^{\lambda_2}(x)$ if either $n < m$, or if $n = m$, and $\lambda_1 \ge \lambda_2$.*

Lemma 10.3.2 leads to

Corollary 10.3.5 *Any TMSR of $F \in \mathcal{F}[a, b]$ induces a complete **representation by topological selections**, namely for any $x \in [a, b]$*

$$F(x) = \bigcup_{n=1}^{N} \{s_n^\lambda(x) : \ s_n^\lambda(x) = \lambda f_n^{low}(x) + (1 - \lambda) f_n^{up}(x), \ \lambda \in [0, 1]\}.$$

$$(10.14)$$

The topological selections in case of a segment function are identical to the generalized Steiner selections (see Section 3.2).

It is also possible to obtain a Castaing representation by topological selections,

$$F(x) = \mathrm{cl}\left(\bigcup_{n=1}^{N} \{s_n^\lambda(x) : \ \lambda \in D\} \right), \quad x \in [a, b], \qquad (10.15)$$

where D is any countable dense set in $[0, 1]$.

10.4. Regularity of SVFs Based on MSRs

In Section 3.3 regularity measures based on representations are derived. This can also be done for MSRs. Since any MSR is determined by its MS-boundaries, and the number of these functions is finite, the regularity measures in this case are simpler.

Any multi-segmental representation of $F \in \mathcal{F}[a, b]$,

$$\mathcal{R} = \{F_n = [f_n^{low}, f_n^{up}], \ n = 1, \ldots, N\},$$

induces a modulus of continuity

$$\omega^{\mathcal{R}}_{[a,b]}(F, \delta) = \max_{1 \le n \le N}\{\omega_{[a,b]}(F_n, \delta)\}. \tag{10.16}$$

Since for a segment function $F(x) = [f^{low}(x), f^{up}(x)]$

$$\omega_{[a,b]}(F, \delta) = \max\{\omega_{[a,b]}(f^{low}, \delta), \omega_{[a,b]}(f^{up}, \delta)\},$$

(10.16) becomes

$$\omega^{\mathcal{R}}_{[a,b]}(F, \delta) = \max\{\omega_{[a,b]}(f, \delta), \ f \in \mathcal{B}(\mathcal{R}, F)\}. \tag{10.17}$$

By arguments as in the proof of Lemma 3.3.1 and by (10.17) one gets

$$\omega_{[a,b]}(F, \delta) \le \omega^{\mathcal{R}}_{[a,b]}(F, \delta) = \max_{1 \le n \le N}\{\omega_{[a,b]}(f_n^{low}, \delta), \omega_{[a,b]}(f_n^{up}, \delta)\}. \tag{10.18}$$

The observation that $F : [a,b] \to K(\mathbb{R})$ is continuous if there is a continuous MSR of F leads to the following notion of smoothness. A multifunction $F \in \mathcal{F}[a,b]$ is termed $MSR - C^k$ if there exists a multi-segmental representation with C^k MS-boundaries. It can be observed in the examples of Fig. 10.1.2 that the SVF in (a) is at most continuous although the boundaries of its graph are smooth curves in the plane. Yet for the SVF in (b) the smoothness is determined by the smoothness of the boundaries of its graph, since they are natural MS-boundaries.

If $\mathcal{R}$ is a topological MSR, in addition to the bounds in (10.18), we have by Corollary 10.2.8

$$\omega_{[a,b]}(f, \delta) \le \omega_{[a,b]}(v_F, \delta), \quad f \in \mathcal{B}(\mathcal{R}, F)$$

and that if $F \in Lip([a,b], L)$ then also $f \in Lip([a,b], L)$.

Thus in case $\mathcal{R}$ is a TMSR there is the tight relation

$$\omega_{[a,b]}(F, \delta) \le \omega^{\mathcal{R}}_{[a,b]}(F, \delta) \le \omega_{[a,b]}(v_F, \delta). \tag{10.19}$$

Note that the MS-boundaries of a TMSR are only continuous, and for most SFVs are not smooth.

The existence of MS-boundaries, which are C^1, is possible for functions in $\mathcal{F}[a, b]$ with holes having the "shape of an eye", namely for any $H \in \mathcal{H}(F)$

$$b_H^{low}, b_H^{up} \in C^1(\Delta_H), \ (b_H^{low})'(x) = (b_H^{up})'(x), \quad x \in \{x_H^l, x_H^r\}. \tag{10.20}$$

A multifunction F with $f^{low}, f^{up} \in C^1([a,b])$ and holes satisfying (10.20) is $MSR - C^1$.

Conditions (10.20) can be straightforwardly extended to guarantee $MSR - C^k$ smoothness.

Chapter 11

Methods Based on Topological Representation

In this chapter we use the topological representations of the previous chapter for the adaptation of approximation operators defined on real-valued functions to SVFs in $\mathcal{F}[a, b]$.

The simplicity of the multifunctions in $\mathcal{F}[a, b]$, allows us to reduce the approximation of such a multifunction to that of a finite number of real-valued functions, defined by the boundaries of the approximated SVF.

The error bounds for the adapted operators are expressed in terms of the regularity properties of the approximated SVF, and are similar to those in the case of real-valued functions.

11.1. Positive Linear Operators Based on TMSRs

In this section we adapt positive linear operators acting on real-valued functions $f : [a, b] \to \mathbb{R}$ to multifunctions in $\mathcal{F}[a, b]$, based on a given TMSR. First we consider segment functions.

In Example 4.4.1 it is observed that a sample-based positive operator A_χ^{Mink} maps $F(x) = [f^{low}(x), f^{up}(x)]$ to $A_\chi^{Mink} F(x) = [A_\chi f^{low}(x), A_\chi f^{up}(x)]$. We extend this adaptation to any positive operator A (not necessarily sample-based), by

$$A^{\mathcal{R}} F(x) = [A f^{low}(x), A f^{up}(x)], \tag{11.1}$$

with $\mathcal{R}$ the representation (10.5) of F with $N = 1$.

Since A is a positive operator, $A^{\mathcal{R}}F$ is a well-defined segment function. Clearly for a continuous segment function F, $A^{\mathcal{R}}F$ approximates F whenever A approximates continuous real-valued functions.

Next we adapt the operator A to $F \in \mathcal{F}[a,b]$ in a way which is based on a given MSR of F.

Definition 11.1.1 Let $F \in \mathcal{F}[a,b]$, let $\mathcal{R}$ be a given MS-representation of F with $\mathcal{B}(\mathcal{R}, F) = \{f_n^{low}, f_n^{up}, \ n = 1, \ldots, N\}$. We define the application of A to F based on $\mathcal{R}$ by

$$(A^{\mathcal{R}}F)(x) = \bigcup_{n=1}^{N}(AF_n)(x) = \bigcup_{n=1}^{N}[(Af_n^{low})(x), (Af_n^{up})(x)], \quad x \in [a,b].$$

(11.2)

It is easy to conclude from this definition that if A approximates continuous real-valued functions, and if $\mathcal{R}$ is a continuous MS-representation, then $A^{\mathcal{R}}F$ approximates F. This is the content of the next theorem.

Theorem 11.1.2 *Let A be a positive approximation operator. Then for $F \in \mathcal{F}[a,b]$ with a continuous MS-representation $\mathcal{R}$*

$$\text{haus}(A^{\mathcal{R}}F(x), F(x)) \leq \max_{f \in \mathcal{B}(\mathcal{R},F)} |Af(x) - f(x)|.$$

(11.3)

Proof Let $\mathcal{R} = \{F_n, \ n = 1, \ldots, N\}$ be the given MSR of F. For any $y \in F(x)$, there exists $1 \leq i \leq N$ such that $y \in F_i(x)$. Then by the definition of the Hausdorff metric

$$\text{dist}(y, (A^{\mathcal{R}}F)(x)) \leq \text{dist}(y, (A^{\mathcal{R}}F_i)(x))$$

$$\leq \text{haus}(F_i(x), (A^{\mathcal{R}}F_i)(x)).$$

(11.4)

Since F_i, $A^{\mathcal{R}}F_i$ are segment functions, we get from (11.4) in view of (2.1)

$$\text{dist}(y, (A^{\mathcal{R}}F)(x)) \leq \max_{1 \leq i \leq N}\{|f_i^{up}(x) - Af_i^{up}(x)|, |f_i^{low}(x) - Af_i^{low}(x)|\}.$$

Thus

$$\sup_{y \in F(x)} \text{dist}(y, (A^{\mathcal{R}}F)(x)) \leq \max_{f \in \mathcal{B}(\mathcal{R},F)} |f(x) - Af(x)|.$$

(11.5)

Similarly

$$\sup_{y \in (A^{\mathcal{R}}F)(x)} \text{dist}(y, F(x)) \leq \max_{f \in \mathcal{B}(\mathcal{R},F)} |f(x) - Af(x)|.$$

This together with (11.5) leads to (11.3). $\qquad\square$

For A satisfying (1.15), it is possible to obtain from (11.3) bounds on the approximation error in terms of the regularity properties of the functions in $\mathcal{B}(\mathcal{R}, F)$.

As is shown in Section 10.1, every continuous MSR contains all the boundaries of F as parts of its MS-boundaries. Thus a good MS-representation should have MS-boundaries with regularity not worse than the regularity of the boundaries of F. Any TMSR meets this condition, and therefore we adapt A to F as follows,

Definition 11.1.3 For $F \in \mathcal{F}[a,b]$ with $\mathcal{R}^*$ a least TMSR, we define the adaptation of A by $A^{\mathcal{R}^*} F$. In case F has only regular PCTs, this adaptation is unique.

With this definition one can obtain bounds on the approximation error in terms of the regularity of F. It follows directly from Theorem 11.1.2 and Corollary 10.2.8 that

Theorem 11.1.4 *Let A_δ be a positive approximation operator depending on a positive parameter δ and satisfying for any $x \in [a,b]$*

$$|(A_\delta f)(x) - f(x)| \le C\,\omega_{[a,b]}(f, \phi(x,\delta)), \quad f \in C[a,b],$$

with ϕ as in (1.15), and let $\mathcal{R}^*$ be a least TMSR of $F \in \mathcal{F}[a,b]$. Then

$$\mathrm{haus}((A_\delta^{\mathcal{R}^*} F)(x), F(x)) \le C\omega_{[a,b]}(v_F, \phi(x,\delta)), \quad x \in [a,b]. \tag{11.6}$$

In particular, if $F \in Lip([a,b], L)$ then

$$\mathrm{haus}((A_\delta^{\mathcal{R}^*} F)(x), F(x)) \le CL\phi(x,\delta), \quad x \in [a,b].$$

As examples of positive operators we consider the Bernstein polynomial operators and the Schoenberg spline operators, and derive error estimates. We illustrate the operation of these adapted operators on two examples.

11.1.1.   *Bernstein polynomial operators*

For $F \in F \in \mathcal{F}[0,1]$ with a least TMSR $\mathcal{R}^*$, we define an adapted Bernstein operator by

$$B_m^{\mathcal{R}^*} F(x) = \bigcup_{n=1}^{N} [B_m f_n^{low}(x), B_m f_n^{up}(x)], \quad x \in [0,1], \tag{11.7}$$

where $f_n^{low}, f_n^{up}, n = 1, \ldots, N$ are the functions in $\mathcal{B}(\mathcal{R}^*, F)$.

Application of Theorem 11.1.4 and (1.21) yields

Corollary 11.1.5 *Let $F \in \mathcal{F}[0,1]$ and let $\mathcal{R}^*$ be a least TMSR of F. Then*

$$\text{haus}(B_m^{\mathcal{R}^*} F(x), F(x)) \leq C\omega_{[0,1]}(v_F, \sqrt{x(1-x)/m}), \quad x \in [0,1].$$

Moreover, for $F \in Lip\ ([0,1], L)$

$$\text{haus}(B_m^{\mathcal{R}^*} F(x), F(x)) \leq CL\sqrt{x(1-x)/m}, \quad x \in [0,1].$$

Here C is a constant independent of F, m and x.

Remark 11.1.6

(i) Since the coefficient of each sample in the Bernstein operator is positive, and since the functions in $\mathcal{B}(\mathcal{R}^*, F)$ are ordered as in (10.4), for m large enough all the functions in

$$\{B_m f : f \in \mathcal{B}(\mathcal{R}^*, F)\} \tag{11.8}$$

are distinct at every $x \in (0,1)$. Therefore $B_m^{\mathcal{R}^*} F$ in (11.7) has a natural TMSR. Moreover, since all the functions in (11.8) are polynomials, $B_m^{\mathcal{R}^*} F$ is $MSR - C^k$ for any k.

(ii) The claim in Corollary 11.1.5 holds for any TMSR.

To illustrate our approach we present the following example.

Example 11.1.7 Consider $F \in \mathcal{F}[0,1]$ of the form

$$F(x) = \begin{cases} \left[-\frac{1}{2}, \frac{1}{2}\right] \backslash \left[-\sqrt{g(x)}, \sqrt{g(x)}\right], & x \in \left[\frac{1}{4}, \frac{3}{4}\right], \\ \left[-\frac{1}{2}, \frac{1}{2}\right], & \text{otherwise}, \end{cases} \tag{11.9}$$

where $g(x) = \frac{1}{16} - (x - \frac{1}{2})^2$.

This multifunction is depicted in gray on (a), (b), (c) and (d) of Fig. 11.1.8. The MS-boundaries of the unique TMSR of F and the two PCTs of F are shown in black on (a). Forty cross-sections of $B_9^{\mathcal{R}^*} F$, $B_{31}^{\mathcal{R}^*} F$ and $B_{50}^{\mathcal{R}^*} F$, are depicted in black on (b), (c) and (d), respectively. The maximal error is attained at $x_1 = \frac{1}{4}$ and $x_2 = \frac{3}{4}$, which are the abscissas of the PCTs of F.

It can be shown that the MS-boundaries θ_H^{up}, θ_H^{low} are Hölder continuous with exponent $1/2$ at $x = x_i$, $i = 1, 2$, and are Lipschitz continuous away from these points. Thus using Theorem 11.1.2 and (1.22) we get that

$$\text{haus}(B_m F(x_i), F(x_i)) \leq C m^{-\frac{1}{4}}, \quad i = 1, 2,$$

while away from x_1, x_2 the error bound is $C m^{-\frac{1}{2}}$. Here and above C is a generic constant. This explains the slow decay of the error at x_1, x_2, as observed from (b),(c),(d) in the figure.

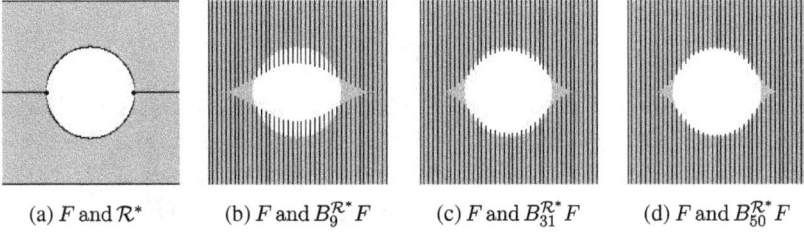

(a) F and $\mathcal{R}^*$ (b) F and $B_9^{\mathcal{R}^*} F$ (c) F and $B_{31}^{\mathcal{R}^*} F$ (d) F and $B_{50}^{\mathcal{R}^*} F$

Figure 11.1.8 F in (11.9) and three Bernstein approximants.

Note that the holes of $B_9^{\mathcal{R}^*} F$, $B_{31}^{\mathcal{R}^*} F$ and $B_{50}^{\mathcal{R}^*} F$ have the shape of an eye in agreement with Remark 11.1.6 and (10.20).

11.1.2. Schoenberg operators

For $F \in \mathcal{F}[0,1]$ with a least TMSR $\mathcal{R}^*$ the adapted Schoenberg spline operators are given by

$$S_{m,N}^{\mathcal{R}^*} F(x) = \bigcup_{n=1}^{N} [S_{m,N} f_n^{low}(x), S_{m,N} f_n^{up}(x)], \quad x \in [0,1]. \qquad (11.10)$$

Then by Theorem 11.1.4 and (1.26) we have

Corollary 11.1.9 *For $F \in \mathcal{F}[0,1]$*

$$\text{haus}(S_{m,N}^{\mathcal{R}^*} F(x), F(x)) \leq \lfloor (m+1)/2 \rfloor \omega_{[0,1]}(v_F, 1/N), \quad x \in [(m-1)/N, 1].$$

Moreover, for $F \in Lip([0,1], L)$

$$\text{haus}(S_{m,N} F(x), F(x)) \leq \lfloor (m+1)/2 \rfloor L \frac{1}{N}, \quad x \in [(m-1)/N, 1].$$

The next example illustrates approximations by the adapted Schoenberg spline operators.

Example 11.1.10 Let the approximated SVF be given by (11.9). F and $S_{3,N}^{\mathcal{R}^*} F$ for $N = 10, 40, 100$ are given in Figure 11.1.11.

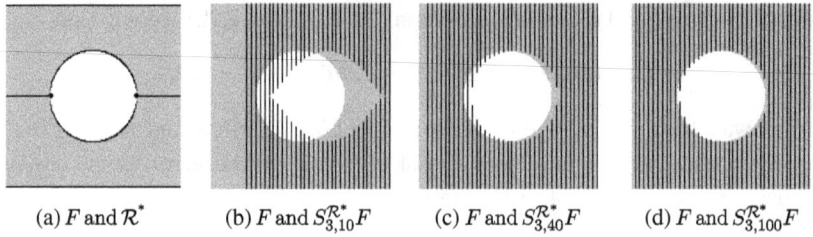

(a) F and $\mathcal{R}^*$ (b) F and $S_{3,10}^{\mathcal{R}^*}F$ (c) F and $S_{3,40}^{\mathcal{R}^*}F$ (d) F and $S_{3,100}^{\mathcal{R}^*}F$

Figure 11.1.11 F in (11.9) and three Schoenberg approximants.

As in the case of Bernstein operators the maximal error is attained at the abscissas of the PCTs of F. By arguments similar to those in the case of Bernstein operators, we obtain that the maximal error decays as $O(N^{-1/2})$, and that the error away from the PCTs of F decays as $O(N^{-1})$.

11.2.   General Operators Based on Topological Selections

In this section we define operators on $F \in \mathcal{F}[a,b]$ using a complete representation by topological selections, similarly to the adaptation of operators based on the complete representation by metric selections, presented in Chapter 8.

Any $F \in \mathcal{F}[a,b]$ has at least one TMSR by the discussion in Section 10.2.1. Thus there exists at least one representation of F by topological selections of the form (10.14).

Definition 11.2.1 Let $F \in \mathcal{F}[a,b]$, let $\mathcal{R}^*$ be a least TMSR of F and let the representation (10.14) be induced by $\mathcal{R}^*$. We adapt to F a linear operator A, acting on continuous real-valued functions, by applying it to the selections in (10.14),

$$A^{\mathcal{R}^*}F(x) = \bigcup_{n=1}^{N}\{(As_n^\lambda)(x) : \lambda \in [0,1]\}, \quad x \in [a,b].$$

Note that in case A is a positive operator, $A^{\mathcal{R}^*}F$ in Definition 11.2.1 is identical to $A^{\mathcal{R}^*}F$ in Definition 11.1.3.

For a linear operator A, not necessarily positive, we have by the linearity and by (10.14)

$$A^{\mathcal{R}^*}F(x) = \bigcup_{n=1}^{N}\{\lambda Af_n^{low}(x) + (1-\lambda)Af_n^{up}(x) : \lambda \in [0,1]\}, \quad x \in [a,b].$$

$$(11.11)$$

The operator $A^{\mathcal{R}^*}F$ given by (11.11) can be simplified to

$$A^{\mathcal{R}^*}F(x) = \bigcup_{n=1}^{N} \mathrm{co}\big(\{Af_n^{low}(x), Af_n^{up}(x)\}\big). \qquad (11.12)$$

This expression for $A^{\mathcal{R}^*}F$ extends the one in (11.2) for positive operators.

It follows from (11.12) that $A^{\mathcal{R}^*}F$ requires the application of A to $2N$ real-valued functions, as in the case of positive operators. Also, the error estimates for $A^{\mathcal{R}^*}F$ in (11.12) are the same as for $A^{\mathcal{R}^*}F$ defined by (11.2). The following theorem extends Theorem 11.1.4.

Theorem 11.2.2 *Let A_δ be a linear approximation operator depending on a positive parameter δ and satisfying for any $x \in [a,b]$*

$$|A_\delta f(x) - f(x)| \leq C\,\omega_{[a,b]}(f, \phi(x,\delta)), \quad f \in C[a,b],$$

with ϕ as in (1.15).

Then for $F \in \mathcal{F}[a,b]])$ and $A_\delta^{\mathcal{R}^*}F$ as in Definition 11.2.1

$$\mathrm{haus}(F(x), A_\delta^{\mathcal{R}^*}F)(x) \leq C\omega_{[a,b]}(v_F, \phi(x,\delta)), \quad x \in [a,b].$$

In particular, if $F \in Lip([a,b], L)$ then

$$\mathrm{haus}(F(x), A_\delta^{\mathcal{R}^*}F(x)) \leq CL\phi(x,\delta), \quad x \in [a,b].$$

Proof The proof is similar to that of Theorem 8.2.2.

By (10.14), for any $y \in F(x)$ there exists $n \in \{1, \ldots, N\}$ and $\lambda \in [0,1]$ such that $y = s_n^\lambda(x)$. Then we have

$$\mathrm{dist}(y, A_\delta^{\mathcal{R}^*}F(x)) = \mathrm{dist}(s_n^\lambda(x), A_\delta^{\mathcal{R}^*}F(x))$$

$$\leq |s_n^\lambda(x) - A_\delta s_n^\lambda(x)| \leq C\omega_{[a,b]}(s_n^\lambda, \phi(x,\delta)). \quad (11.13)$$

Thus

$$\sup_{y \in F(x)} \mathrm{dist}(y, A_\delta^{\mathcal{R}^*}F(x)) \leq C \sup_{\lambda, n} \omega_{[a,b]}(s_n^\lambda, \phi(x,\delta))$$

$$\leq C\omega_{[a,b]}(v_F, \phi(x,\delta))$$

where the last inequality follows from Lemma 10.3.3.

Next, for any $a \in A_\delta^{\mathcal{R}^*} F(x)$ there exists s_m^μ such that $a = A_\delta s_m^\mu(x)$. Then using Lemma 10.3.3 again we obtain

$$\text{dist}(a, F(x)) = \text{dist}(A_\delta s_m^\mu(x), F(x)) \le |s_m^\mu(x) - A_\delta s_m^\mu(x)|$$

$$\le C\omega_{[a,b]}(s_m^\mu, \phi(x, \delta)) \le C\omega_{[a,b]}(v_F, \phi(x, \delta)). \quad (11.14)$$

Combining (11.13) and (11.14) we get the first claim of the theorem.

In case $F \in Lip([a, b], L)$ we can improve (11.13) and (11.14) to obtain the second claim of the theorem. □

Remark 11.2.3

(i) By similar arguments to those leading to Corollary 8.2.3, we get for $F \in Lip([a, b], L)$

$$\text{haus}(F(x), P_\chi^{\mathcal{R}^*} F(x)) \le \frac{C \log N}{N},$$

where P_χ is the polynomial interpolation operator (1.29) at the points χ, which are the roots of the Chebyshev polynomial of degree $N + 1$.

(ii) TMSRs have continuous boundaries which are C^1 only for very special SVFs. Therefore the methods of this chapter provide only low-rate approximations. Yet for F which is $MSR - C^k$ it is possible to get higher order approximation, by the adaptation methods presented in this chapter.

11.3. Bibliographical Notes to Part III

The results in Part III are based on [46, 67].

The function studied in this part of the book can be regarded as extensions of the interval-valued functions (segment functions). The latter functions have been extensively studied in the last half century within the subject of Interval Analysis, motivated by the study of the accumulation of round-off errors in Scientific computing, see, e.g., [1, 28, 68, 69].

Various notions of continuity of segment functions can be found, e.g., in [2, 3].

The result of Theorem 9.2.1 is mentioned in [29], Section 2.3.

Bibliography

[1]   G. Alefeld, G. Mayer, (2000). Interval analysis: theory and applications, *Journal of Computational and Applied Mathematics*, 121, pp. 421–464.

[2]   R. Anguelov, (2007). Algebraic Computations with Hausdorff Continuous Functions, *Serdica Journal of Computing*, 1, pp. 443–454.

[3]   R. Anguelov, S. Markov, B. Sendov, (2006). The set of Hausdorff continuous functions — the largest linear space of interval functions, *Reliable Computing*, 12, pp. 337–363.

[4]   K. J. Arrow, F. H. Hahn, (1971). *General Competitive Analysis*, Holden-Day, San Francisco.

[5]   Z. Artstein, (1989). Piecewise linear approximations of set-valued maps, *Journal of Approximation Theory*, 56, pp. 41–47.

[6]   J.-P. Aubin, (1999). *Mutational and Morphological Analysis*, Birkhäuser Boston Inc., Boston.

[7]   J.-P. Aubin, A. Cellina, (1984). *Differential Inclusions*, Springer-Verlag, Berlin.

[8]   J.-P. Aubin, H. Frankowska, (1990). *Set-Valued Analysis*, Birkhäuser Boston Inc., Boston.

[9]   R. Baier, (1995). Mengenwertige Integration und die diskrete Approximation erreichbarer Mengen, Dissertation, Universität Bayreuth, Bayreuth, *Bayreuther Mathematische Schriften*, 50.

[10] R. Baier, N. Dyn, E. Farkhi, (2002). Metric Averages of One Dimensional Compact Sets, in: C. Chui, L. Schumaker and J. Stoeckler (eds), *Approximation theory X*, Vanderbilt University Press, Nashville, pp. 9–22.

[11] R. Baier, E. Farkhi, (2001). Directed derivatives of convex compact-valued mappings, in: N. Hadjisavvas, P. Pardalos (eds), *Advances in Convex Analysis and Global Optimization*. Nonconvex Optimization and Its Applications, 54, Kluwer, Dordrecht, pp. 501–514.

[12] R. Baier, E. Farkhi, (2001). Differences of convex compact sets in the space of directed sets, Part I: The space of directed sets. *Set-Valued Analysis*, 9, pp. 217–245.

145

[13] R. Baier, E. Farkhi, (2001). Differences of convex compact sets in the space of directed sets, Part II: Visualization of directed sets, *Set-Valued Analysis*, 9, pp. 247–272.

[14] R. Baier, E. Farkhi, (2007). Regularity and integration of set-valued maps represented by generalized Steiner points. *Set-Valued Analysis*, 15, pp. 185–207.

[15] R. Baier, F. Lempio, (1994). Approximating reachable sets by extrapolation methods, in: P. J. Laurent, A. Le Méhauteé and L. L. Schumaker (eds), *Curves and Surfaces in Geometric Design*, A. K. Peters, Wellesley, pp. 9–18.

[16] R. Baier, F. Lempio, (1994). Computing Aumann's integral, in: A. B. Kurzhanski, V. M. Veliov (eds), *Modeling Techniques for Uncertain Systems*, Progress in Systems and Control Theory, 18, Birkhäuser, Boston, pp. 7–92.

[17] R. Baier, G. Perria, (2011). Set-valued Hermite interpolation, *Journal of Approximation Theory*, 163, pp. 1349–1372.

[18] C. de Boor, (2001). *A Practical Guide to Splines*, Springer-Verlag, New York.

[19] D. Burago, Y. Burago, S. Ivanov, (2001). *A Course in Metric Geometry*, American Mathematical Society, Providence.

[20] J. W. S. Cassels, (1975). Measures of the non-convexity of sets and the Shapley–Folkman–Starr theorem, *Mathematical Proceedings of the Cambridge Philosophical Society*, 78, pp. 433–436.

[21] C. Castaing, M. Valadier, (1977). *Convex Analysis and Measurable Multifunctions*, Lecture Notes in Math., 580, Springer-Verlag, Berlin.

[22] A. Cavaretta, W. Dahmen, C. Micchelli, (1991). *Stationary Subdivision*, Memoirs of the AMS, 453, American Mathematical Society, Providence.

[23] V. V. Chistyakov, (1997). On mappings of bounded variation, *Journal of Dynamical and Control Systems*, 3, pp. 261–289.

[24] V. V. Chistyakov, (2004), Selections of bounded variation. *Journal of Applied Analysis*, 10, pp. 1–82.

[25] V. Chistyakov, D. Repovš, (2007). Selections of bounded variation under the excess restrictions. *Journal of Mathematical Analysis and Applications*, 331, pp. 873–885.

[26] F. Clarke, Y. Ledyaev, R. Stern, P. Wolenski, (1998). *Nonsmooth Analysis and Control Theory*, Graduate Texts in Mathematics, 178, Springer-Verlag, New York.

[27] B. Csebfalvi, L. Neumann, A. Kanitsar, E. Groller, (2002). Smooth shape-based interpolation using the conjugate gradient method, *Proceedings of Vision, Modeling, and Visualization*, pp. 123–130.

[28] H. Dawood, (2011). *Theories of Interval Arithmetic: Mathematical Foundations and Applications*, Lambert Academic Publishing, Saarbrücken.

[29] K. Deimling, (1992). *Multivalued Differential Equations*, W. De Gruyter, Berlin.

[30] D. Dentcheva, (1998). Differentiable selections and Castaing representations of multifunctions, *Journal of Mathematical Analysis and Applications*, 223(2), pp. 371–396.

[31] R. DeVore, G. Lorentz, (1993). *Constructive Approximation*, Springer-Verlag, Berlin.

[32] R. DeVore, G. Petrova, P. Wojtaszczyk, (2008). Anisotropic smoothness spaces via level sets, *Communications on Pure and Applied Mathematics*, 61, pp. 1264–1297.

[33] G. Dommisch, (1987). On the existence of Lipschitz-continuous and differentiable selections for multifunctions, in: J. Guddat, H. Jongen, B. Kummer and F. Nožička (eds), *Parametric optimization and related topics*, Mathematical Research, 35, Akademie-Verlag, Berlin, pp. 60–73.

[34] T. Donchev, E. Farkhi, (1990). Moduli of smoothness of vector-valued functions of a real variable and applications, *Numerical Functional Analysis and Optimization*, 11 (5,6), pp. 497–509.

[35] T. Donchev, E. Farkhi, (1999). Approximations of one-sided Lipschitz differential inclusions with discontinuous right-hand sides, in: A. Ioffe, S. Reich and I. Shafrir (eds), *Calculus of Variations and Differential Equations*, Research Notes in Mathematics, 410, CRC Press, Boca Raton, pp. 101–118.

[36] A. L. Dontchev, E. Farkhi, (1988). An averaged modulus of continuity for multivalued maps and its applications to differential inclusions, in: B. Sendov, P. Petrushev, K. Ivanov, R. Maleev (eds), *Constructive theory of functions*, Publishing House of the Bulgarian Academy of Sciences, Sofia, pp. 127–131.

[37] A. L. Dontchev, E. Farkhi, (1989). Error estimates for discretized differential inclusions, *Computing*, 41, pp. 349–358.

[38] A. L. Dontchev, F. Lempio, (1992). Difference Methods for Differential Inclusions: A Survey, *SIAM Review*, 34, pp. 263–294.

[39] A. L. Dontchev, M. P. Polis, V. M. Veliov, (2000). A dual method for parameter identification under deterministic uncertainty, *IEEE Transactions on Automatic Control*, 45 (7), pp. 1341–1346.

[40] N. Dyn, (1992). Subdivision schemes in Computer-Aided Geometric Design, in: W. Light (ed), *Advances in Numerical Analysis: Wavelets, Subdivision Algorithms and Radial Basis Functions*, Clarendon Press, Oxford, pp. 36–104.

[41] N. Dyn, E. Farkhi, (2000). Spline subdivision schemes for convex compact sets, *Journal of Computational and Applied Mathematics*, 119, pp. 133–144.

[42] N. Dyn, E. Farkhi, (2001). Spline Subdivision Schemes for Compact Sets with metric averages, in: K. Kopotun, T. Lyche and M. Neamtu (eds), *Trends in Approximation Theory*, Vanderbilt University Press, Nashville, pp. 95–104.

[43] N. Dyn, E. Farkhi, (2004). Set-valued approximations with Minkowski averages — convergence and convexification rates, *Numerical Functional Analysis and Optimization*, 25, pp. 363–377.

[44] N. Dyn, E. Farkhi, A. Mokhov, (2007). Approximations of set-valued functions by metric linear operators, *Constructive Approximation*, 25, pp. 193–209.

[45] N. Dyn, E. Farkhi, A. Mokhov, (2007). Approximation of univariate set-valued functions — an overview, *Serdica Mathematical Journal* 33, pp. 495–514.

[46] N. Dyn, E. Farkhi, A. Mokhov, (2009). Multi-segmental representations and approximation of set-valued functions with 1D images, *Journal of Approximation Theory*, 159, pp. 39–60.

[47] N. Dyn, D. Levin, (2002). Subdivision schemes in geometric modelling, *Acta Numerica*, 11, pp. 73–144.

[48] N. Dyn, A. Mokhov, (2006). Approximations of set-valued functions based on the metric average, *Rendiconti di Matematica e delle sue Applicazioni. Serie VII*, 26, pp. 249–266.

[49] G. Freud, (1967). On approximation by positive linear methods, Part I, *Studia Scientiarum Mathematicarum Hungarica*, 2, pp. 63–66.

[50] B. V. Gnedenko, (1963). *The Theory of Probability*, Chelsea, New York.

[51] C. Le Guernic, A. Girard, (2010). Reachability analysis of linear systems using support functions, *Nonlinear Analysis. Hybrid Systems*, 4, pp. 250–262.

[52] G. Herman, J. Zheng, C. Bucholtz, (1992). Shape-Based Interpolation, *IEEE Computer Graphics and Applications-CGA*, 12 (3), pp. 69–79.

[53] H. Hermes, (1971). On continuous and measurable selections and the existence of solutions of generalized differential equations, *Proceedings of AMS*, 29, pp. 535–542.

[54] P. L. Hörmander, (1954). Sur la fonction d'appui des ensembles convexes dans un espace localement convexe, *Arkiv för Matematik*, 3, pp. 181–186.

[55] S. Kels, N. Dyn, E. Lipovetsky, (2010). Computation of the metric average of 2D sets with piecewise linear boundaries, *Algorithms (Basel)*, 3 (3), pp. 265–275.

[56] S. Kels, N. Dyn, (2011). Reconstruction of 3D objects from 2D cross-sections with the 4-point subdivision scheme adapted to sets, *Computers & Graphics*, 35, pp. 741–746.

[57] S. Kels, N. Dyn, (2013). Subdivision schemes of sets and the approximation of set-valued functions in the symmetric difference metric, *Foundations of Computational Mathematics*, 13, pp. 835–865.

[58] S. Kels, N. Dyn, (2014). Bernstein-type approximation of set-valued functions in the symmetric difference metric, *Discrete and Continuous Dynamical Systems — Series A*, 34, pp. 1041–1060.

[59] D. Klatte, B. Kummer, (2002). *Nonsmooth Equations in Optimization: Regularity, Calculus Methods and Applications*, Kluwer, New York.

[60] V. Klee, (1965). A theorem on convex kernels, *Mathematika*, 12, pp. 89–93.

[61] P. P. Korovkin, (1959). *Linear operators and approximation theory*. Russian Monographs and Texts on Advanced Mathematics and Physics, Vol. III, Gordon and Breach Publishers, Inc., New York. Translated from the Russian (1960), *Lineinye operatory i teoria priblizhenii*, Moskva, Fizmatgiz.

[62] J. Lane, R. Riesenfeld, (1980). A theoretical development for the computer generation and display of piecewise polynomial surfaces, *IEEE transactions on Pattern Analysis and Machine Intelligence*, 1, pp. 35–46.

[63] F. Lempio, (1995). Set-valued interpolation, differential inclusions, and sensitivity in optimization, in: R. Lucchetti, J. Revalski (eds), *Recent Developments in Well-Posed Variational Problems*, Mathematics and Its Applications, 331, Kluwer Academic Publishers, Dordrecht, pp. 137–169.

[64] R. Levie, (2014). Line Cross-Section Models: Hybrid volume-surface models of 3D objects based on 1D cross-sections, MSc thesis, Tel-Aviv University.

[65] D. Levin, (1986). Multidimensional reconstruction by set-valued approximations, *IMA Journal of Numerical Analysis* 6 , pp. 173–184.

[66] E. Lipovetsky, N. Dyn, (2007). An efficient algorithm for the computation of the metric average of two intersecting convex polygons, with application to morphing. *Advances in Computational Mathematics*, 26 (1–3), pp. 269–282.

[67] A. Mokhov, (2011). Approximation and Representation of Set-Valued functions with Compact Images, PhD Thesis, Tel-Aviv University.

[68] R. E. Moore, (1966). *Interval analysis*, Prentice-Hall, Inc., Englewood Cliffs.

[69] R. E. Moore, R. B. Kearfott, M. J. Cloud, (2009). *Introduction to Interval Analysis*, SIAM Press, Philadelphia.

[70] B. Mordukhovich, (1995). Discrete approximations and refined Euler-Lagrange conditions for nonconvex differential inclusions, *SIAM Journal on Control and Optimization*, 33, pp. 882–915.

[71] B. Mordukhovich, (2005). *Variational Analysis and Generalized Differentiation I: Basic Theory*, Springer-Verlag, Berlin.

[72] B. Mordukhovich, (2005). *Variational Analysis and Generalized Differentiation II: Applications*, Springer-Verlag, Berlin.

[73] M. Mureşan, (2010). Set-valued approximation of multifunctions, *Studia. Universitatis Babeş-Bolyai Mathematica*, 55 (1), pp. 107–148.

[74] S. B. Myers, (1945) Arcs and Geodesics in Metric Spaces, *Transactions of the AMS*, 57, pp. 217–227.

[75] I. P. Natanson, (1965). *Constructive Function Theory*, Frederick Ungar, New York.

[76] M. S. Nikolskii, (1989). Approximation of convex-valued continuous multi-valued mappings, *Soviet Mathematics Doklady*, 40, pp. 406-409. Translated from the Russian (1990) *Doklady Akademii Nauk SSSR*, 308, pp. 1047–1050.

[77] M. S. Nikolskii, (1990). Approximation of a continuous multivalued mapping by constant multivalued mappings, *Moscow University Computational Mathematics and Cybernetics*, 1, pp. 73–76. Translated from the Russian *Vestnik Moskovskogo Universiteta. Seriya XV. Vychislitelnaya Matematika i Kibernetika*, 1, pp. 76–79, 82.

[78] M. S. Nikolskii, (1990). Approximation of continuous convex-valued multi-valued mappings, (Russian), *Optimization*, 21 (2), pp. 209–214.

[79] G. Perria, (2007). Set-Valued Interpolation, Dissertation, Universität Bayreuth, Bayreuth, *Bayreuther Mathematische Schriften*, 79.

[80] H. Prautzsch, W. Boehm, M. Paluszny, (2002). *Bezier and B-Spline Techniques*, Springer-Verlag, Berlin.

[81] H. Rådström, (1952). An embedding theorem for spaces of convex sets. *Proceedings of the American Mathematical Society*, 3, pp. 165–169.

[82] R. T. Rockafellar, (1970). *Convex Analysis*, Princeton University Press, Princeton.

[83] R. T. Rockafellar, R. Wets, (1998). *Variational Analysis*, Springer-Verlag, Berlin.

[84] R. Schneider, (1993). *Convex Bodies: The Brunn–Minkowski Theory*, Cambridge University Press, Cambridge.

[85] B. Sendov, V. Popov, (1988). *The averaged moduli of smoothness: applications in numerical methods and approximation*, Wiley, Chichester.

[86] J.-P. Serra, P. Soille (eds), (1994). *Mathematical morphology and its applications to image processing*, Kluwer, Dordrecht.

[87] D. Silin, (1997). On set-valued differentiation and integration, *Set-Valued Analysis*, 5 (2), pp. 107–146.

[88] R. Starr, (1969). Quasi-equilibria in markets with non-convex preferences, *Econometrica*, 37, pp. 25–38.

[89] F. A. Toranzos, (1967). Radial Functions of Convex and Star-Shaped Bodies, *The American Mathematical Monthly*, 74, pp. 278–280.

[90] V. Veliov, (1992). Second-order discrete approximation to linear differential inclusions, *SIAM Journal on Numerical Analysis*, 29 (2), pp. 439–451.

[91] R. A. Vitale, (1979). Approximations of convex set-valued functions, *Journal of Approximation Theory*, 26, pp. 301–316.

[92] J. Wallner, N. Dyn, (2005). Convergence and C^1 analysis of subdividision schemes on manifolds by proximity, *CAGD*, 22, pp. 593–622.

[93] R. Wegmann, (1980). Einige Masszahlen für nichtkonvexe Mengen, *Archiv der Mathematik*, 34, pp. 69–74.

[94] Y. Yang, Z. Wu, (2005). Subdivision schemes for non-convex compact sets with a new defifnion of set average, preprint, *IMS Preprint Series*, National University of Syngapore, 15.

[95] Q. J. Zhu, N. Zhang, Y. He, (1992). Algorithm for determining the reachability set of a linear control system, *Journal of Optimization Theory and Applications*, 72 (2), pp. 333–353.

Index

www.ingramcontent.com/pod-product-compliance
Lightning Source LLC
Chambersburg PA
CBHW050630190326
41458CB00008B/2206